Melatonin - The Hormone of Darkness and its Therapeutic Potential and Perspectives

Edited by Marilena Vlachou

Published in London, United Kingdom

IntechOpen

Supporting open minds since 2005

Melatonin - The Hormone of Darkness and its Therapeutic Potential and Perspectives
http://dx.doi.org/10.5772/intechopen.80180
Edited by Marilena Vlachou

Part of IntechOpen Book Series: Physiology, Volume 9
Book Series Editor: Angel Catala

Contributors
Emiliano Raúl Diez, Natalia Jorgelina Prado, Esther Pueyo, Margarita Segovia-Roldan, Biao Gong, Yanyan Yan, Qinghua Shi, Alexander Zakharov, Elena Khivintseva, Ioulia K. Tseti, Akikazu Takada, Natassa Pippa, Costas Demetzos, Angeliki Siamidi, Yannis Dotsikas, Andrew Tsotinis, Ioannis P. Papanastasiou, Hiroi Tomioka, Junichi Masuda, Akira Iwanami

First published in London, United Kingdom, 2020 by IntechOpen
IntechOpen is the global imprint of INTECHOPEN LIMITED, registered in England and Wales, registration number: 11086078, 7th floor, 10 Lower Thames Street, London, EC3R 6AF, United Kingdom
Printed in Croatia

British Library Cataloguing-in-Publication Data
A catalogue record for this book is available from the British Library

Additional hard and PDF copies can be obtained from orders@intechopen.com

Melatonin - The Hormone of Darkness and its Therapeutic Potential and Perspectives
Edited by Marilena Vlachou
p. cm.
Print ISBN 978-1-83962-908-2
Online ISBN 978-1-83962-909-9
eBook (PDF) ISBN 978-1-83962-910-5
ISSN 2631-8261

For EU product safety concerns:
IN TECH d.o.o., Prolaz Marije Krucifikse Kozulić 3, 51000 Rijeka, Croatia, info@intechopen.com or visit our website at intechopen.com.

IntechOpen Book Series
Physiology
Volume 9

Marilena Vlachou is an Assistant Professor at the National and Kapodistrian University of Athens (NKUoA), Greece. She obtained her Pharmacy and PhD (Pharmaceutical Technology) degrees from the NKUoA. Just prior to obtaining her PhD she moved to the University of Rhode Island, United States, as a Visiting Research Scientist to conduct state-of-the art research on Pharmaceutical Technology techniques. Her research interests include: the formulation and *in vitro* release of bioactive substances from topical formulations; the efficacy and safety of formulations in skin disease therapies; the modified release of novel synthetic derivatives, with diverse activity; the investigation of the physicochemical properties of new excipients, including those of marine origin (Ulvans), and nanomaterials, with respect to their interaction with active pharmaceutical ingredients (APIs).

Editor of Volume 9:
Ass. Professor Marilena Vlachou
National and Kapodistrian University of Athens, Department of Pharmacy, Greece

Book Series Editor: Angel Catala
National University of La Plata, Argentina

Scope of the Series

Modern physiology requires a comprehensive understanding of the integration of tissues and organs throughout the mammalian body, including the expression, structure, and function of molecular and cellular components. While a daunting task, learning is facilitated by our identification of common, effective signaling pathways employed by nature to sustain life. As a main example, the cellular interplay between intracellular Ca2 increases and changes in plasma membrane potential is integral to coordinating blood flow, governing the exocytosis of neurotransmitters and modulating genetic expression. Further, in this manner, understanding the systemic interplay between the cardiovascular and nervous systems has now become more important than ever as human populations age and mechanisms of cellular oxidative signaling are utilized for sustaining life. Altogether, physiological research enables our identification of clear and precise points of transition from health to development of multi-morbidity during the inevitable aging process (e.g., diabetes, hypertension, chronic kidney disease, heart failure, age-related macular degeneration; cancer). With consideration of all organ systems (e.g., brain, heart, lung, liver; gut, kidney, eye) and the interactions thereof, this Physiology Series will address aims of resolve (1) Aging physiology and progress of chronic diseases (2) Examination of key cellular pathways as they relate to calcium, oxidative stress, and electrical signaling & (3) how changes in plasma membrane produced by lipid peroxidation products affects aging physiology.

Contents

Preface

The pineal hormone of "darkness," melatonin (*N*-acetyl-5-methoxytryptamine), is biosynthesized and secreted mainly at night. It is an important component in the regulation of seasonal and circadian rhythms. Its action is believed to be mediated through a family of specific, high-affinity, G-protein-coupled cell membrane receptors. It is ubiquitous throughout both the animal and plant kingdoms and must have a long evolutionary history as a hormone. The secretion of the hormone is closely synchronized with the habitual hours of sleep in humans. Ingestion of melatonin affects sleep propensity as well as duration and quality of sleep. Sleep problems become more common in the elderly in whom there is also a loss in the production of melatonin.

Melatonin treatment has therapeutic value in some blind subjects, restoring their disturbed circadian rhythm. It also has potential in the treatment of seasonal affective disorder (SAD), which afflicts some people during the short winter days, and it has been used to reset the clock in sufferers from jet lag. Melatonin has been implicated in a range of other conditions, including Parkinson's disease, Alzheimer's and other neurodegenerative conditions, and in certain cancers. It has also recently been shown that melatonin can be synthesized by mammalian skin, where it may be important in regulating hair growth and pigmentation physiology.

Several synthetic molecules have shown interesting melatoninergic activity, frequently greater and more selective than that of the endogenous hormone. Ramelteon was the first prescription medication for insomnia, and the only hypnotic indicated for long-term treatment of insomnia as it does not have hangover, addictive, or withdrawal effects. Another melatonin analogue, agomelatine, has recently been introduced as an antidepressant and appears to have few side effects.

The use of melatonin as a drug is hampered by its short biological half-life and poor bioavailability. As a result, dosage forms, which mimic the physiologically secreted melatonin concentration versus time model, are limited.

The eight chapters in this book deal with all these aspects. The initial chapter provides an account of most of the synthetic melatoninergic agents available thus far, and is addressed to a wide non-cognizant readership. Briefly, this chapter focuses on the synthetic routes towards synthetic melatonin derivatives, first of their aromatic nucleus, then of the functionalities that have been introduced to the nucleus, and finally those analogues with restrained conformations and those that are optically active. The second chapter is pivotal to the research on melatonin's direct involvement with sleep and the dysfunctions caused by insomnia, which are also presented in the third chapter. Moreover, in the latter chapter, the use of synthetic melatonin, as a food supplement in various dosage forms such as pills, granules for oral solution, orodispersable granules, and syrups in order to address patients' needs, is described. The fourth chapter focuses on the electrophysiological and the antiarrhythmic properties of melatonin. The acute and chronic protective mechanisms of melatonin are analyzed with an emphasis on transmembrane potentials and intercellular communication. An outstanding antifibrilatory effect is claimed to make melatonin a novel antiarrhythmic agent worthy of further exploration in the path to clinical applications. In the context of the fifth chapter, a review of the related literature on the modified release of melatonin from its *per os* administered formulations is presented, including utilization of design of experiments (DoE) for the selection of the optimal composition of melatonin formulations. The chapter offers an account of the recent advantages on the hormone's solid dosage forms suitable for treating sleep disorders, referring either to its onset or maintenance. The sixth chapter describes the conventional solid and liquid forms (i.e., tables, capsules, suspensions, etc.) and the nanoformulations (i.e., liposomes, niosomes, polymeric nanoparticles, chitosomes, calcium alginate beads, etc.) of melatonin, focusing on its release kinetics from pharmaceutical vehicles. These systems have been designed and developed as platforms for the delivery and release of melatonin. The penultimate chapter describes melatonin's physiological functions, including regulating plant growth, promoting seed germination, controlling root development and delaying leaf senescence. The final chapter refers to the active transport of 5-HT by platelets, which has been shown to be significantly correlated with melatonin blood levels. This expounds either a direct effect of melatonin on 5-HT active transport or the influence of the suprachiasmatic nucleus on serotonin uptake by platelets, pathways that are considered to be associated with various neurobiological alterations, like hyperactivity of the hypothalamic-pituitary-adrenal axis, altered neuroplasticity, and altered circadian rhythms.

Acknowledgements. I would like to thank all the authors, who contributed to the compilation of this text: Professor Andrew Tsotinis and Ass. Professor Ioannis Papanastasiou (Chapter 1); Dr. Alexander Zakharov (Chapter 2); Ms. Ioulia K. Tseti, MSc (Chapter 3); Ms. Natalia Jorgelina Prado, Dr. Margarita Segovia-Roldan, Dr. Emiliano Raúl Diez, Professor Esther Pueyo (Chapter 4); Dr. Angeliki Siamidi and Ass. Professor Yannis Dotsikas (Chapter 5); Dr. Natassa Pippa and Professor Costas Demetzos (Chapter 6); Yanyan Yan, Professor Qinghua Shi, Professor Biao Gong (Chapter 7); Dr. Hiroi Tomioka, Junichi Masuda MD, Dr. Akikazu Takada, Ms. Akira Iwanami (Chapter 8).

Marilena Vlachou
Assistant Professor,
Department of Pharmacy,
Section of Pharmaceutical Technology,
National and Kapodistrian University of Athens,
Panepistimioupoli-Zografou,
Athens, Greece

Section 1

Melatonin Receptor Agonists and Antagonists

Chapter 1

Synthetic Melatonin Receptor Agonists and Antagonists

Andrew Tsotinis and Ioannis P. Papanastasiou

Abstract

The functions of the pineal hormone melatonin are of intense and continuous interest. Synthetic melatonin receptor analogues, as agonists and antagonists, have been explored, and the molecule can be viewed as consisting of an indole nucleus, acting mainly as a spacer, and the C5-OMe and the C3-ethylamido side chains, acting as pharmacophoric components. The present chapter focuses on the synthetic routes towards these melatonin derivatives, first the aromatic nucleus, then the functionalities that have been introduced to the nucleus, and finally those analogues with restrained conformations and those that are optically active. The importance of the various parameters involved in the agonist and antagonist profile of the compounds is indicated, as is the difference in the action of the chiral melatoninergics.

Keywords: melatonin, indole and bioisosteric derivatives, constrained polycyclic analogues, chiral melatonin analogues

1. Introduction

Melatonin (*N*-acetyl-5-methoxytryptamine **1**) is a hormone ubiquitously distributed in a variety of organisms, such as bacteria, unicellular algae, fungi, plants, vertebrates, and mammalians [1]. Melatonin is mainly known to regulate circadian rhythms by synchronization to environmental cues but participates also in diverse important physiological processes, such as regulation of the visual functions, glucose metabolism, and immune functions (**Figure 1**) [2]. The functions of melatonin are modulated through its binding to G protein-coupled receptors (GPCRs), which activate signaling pathways, as a cascade effect [3]. Up to date, two different types of melatonin receptors have been described in mammals: type 1A (MT1) and type 1B (MT2). Both receptors are located in many regions in the central nervous system and in peripheral tissues as well [4]. X-ray free electron laser (XFEL) studies have recently revealed that MT1 binding site is extremely compact, and ligands interact with MT1 mainly by strong aromatic stacking with Phe179 and auxiliary hydrogen bonds with Asn162 and Gln181 [5]. Comparison of the structures of MT2 and MT1 indicated that, despite conservation of the orthosteric ligand binding site residues, there are significant conformational variations between both melatonin receptor subtypes, which justify the selectivity between the two subtypes [6]. Melatonin was proven to bind to one more co-substrate binding site (MT3), which is a quinone reductase-2 [7]. Melatonin receptors had been cloned in 1990s [8–10] but characterized and described in the 1980s by using the radiolabeled 2-[^{125}I]-iodomelatonin and ^{3}H-melatonin ligands [11, 12]. Herein, we are reviewing the synthetic routes of

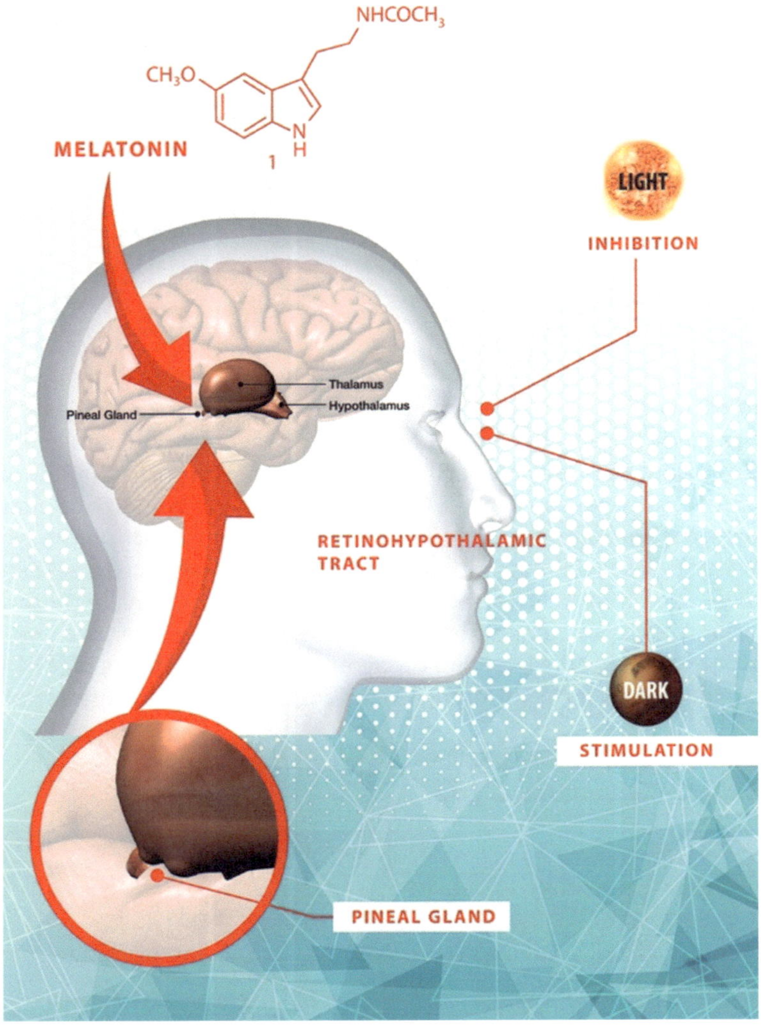

Figure 1.
Regulation of melatonin production.

the main indole and bioisosteric aromatic nucleus derivatives: first, the conformationally restricted; the active chiral compounds second; and the derivatives with substituted 3-side chains third.

2. Indole and bioisosteric derivatives

A guide of general principles has been applied throughout SARs for both melatonin receptors. The C5-OMe group of the indole ring is optimal, while the same substituent at positions 4, 6, or 7 leads to a drastic loss of affinity. However, congeners with a halogen at the 5-position do retain high affinity [13]. The relative position of the methoxy group and the *N*-acetylaminoethyl side chain seems to be the most important structural feature that increases the melatonin receptor binding affinity [14–16]. The syntheses of these derivatives are based on classic chemical procedures [17–19]. The indole ring could also be considered as a spacer [20, 21] with the pyrrole portion not involved in the receptor binding pocket, because it can be replaced by diverse aromatic scaffolds, such as naphthalene, benzofuran, benzothiophene, or benzocycloalkane rings [14, 22, 23]. Various congeners with substitutions in the positions 2 and 6 of melatonin have been synthesized. Substituents, like methyl, phenyl, or halogen at position 2 of melatonin, can increase receptor binding affinity by *ca* tenfold [24–27]. The presence of an optimal *N*-acyl group with a 2-halogen substitution exhibits very potent affinity [28].

Interestingly, substituents on the 2-position seem to direct the *N*-acetylaminoethyl side chain into the optimal conformation for interaction with the receptor and increase the ligand affinity [29, 30]. 6-Substituted analogues have been prepared [31] with the aim of retarding metabolism, because melatonin is degraded rapidly in vivo, mainly in the liver, by 6-hydroxylation followed by conjugation and excretion in the urine. A halogen substituent at the 6-position reduces binding affinity nonsignificantly, while the binding affinity of 6-hydroxymelatonin is decreased by 5 to 10 times and 6-methoxymelatonin by more than 100 times [32].

One of the synthetic routes for the production of 5-methoxyindole (**4**) is via the Leimgruber-Batcho reaction [33], modified by Repke and Ferguson [34] (**Figure 2**). A successful side chain functionalization was reported by Ates-Alagoz et al. [35] using the Vilsmeier-Haack formylation reaction of 5-methoxyindole (**4**). On the other hand, Righi et al. [36] applied the direct C3 reductive alkylation of *N*-benzyl-5-methoxyindole (**8**), as described in **Figure 2**.

In an attempt to map the receptor requirements, a series of phenylalkyl amides **9–11** were prepared and proven to exhibit the minimal structure required for the ligand recognition by melatonin receptors [16, 37, 38] (**Figure 3**).

Some C3-modified melatonin analogues have exhibited interesting melatoninergic activities. It has been shown that small modifications in the acyl chain are able to change the binding affinity for melatonin receptors. A typical modification to increase the activity is the replacement of the acetyl by an *N*-butanoyl chain. Depreux et al. reported a 100-fold higher affinity of 5-methoxy-*N*-butanoyltryptamine than melatonin [14]. Tsotinis et al. reported that upon the appropriate functionalization at the end of the C2 side chain, the azido compounds **16** were produced, which serve as photoactivity labels, while the respective isothiocyanate compounds **17** serve as electrophilic probes (**Figure 4**), in order to produce adducts covalently linked to key amino acid residues of the melatonin receptor subtypes [39].

Luzindole, N-acetyl-2-benzyltryptamine (**21**), is a selective melatonin receptor antagonist with approximately 11- to 25-fold higher affinity for the MT2 than the

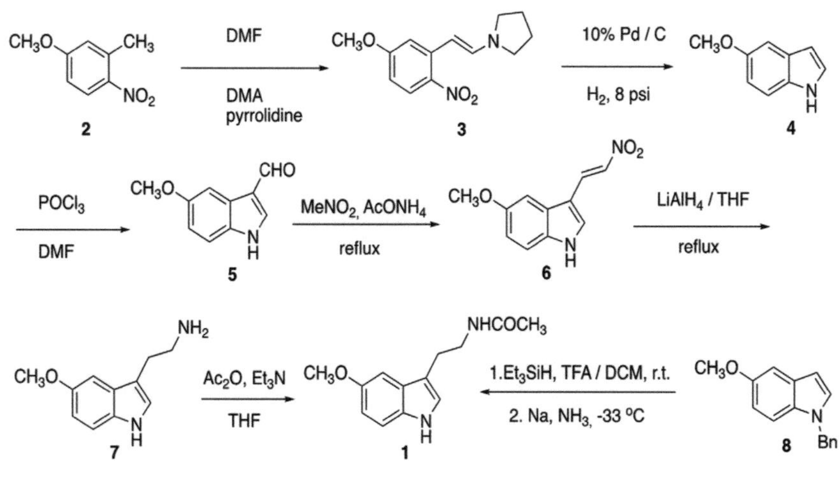

Figure 2.
Highlighted synthetic routes of melatonin 1.

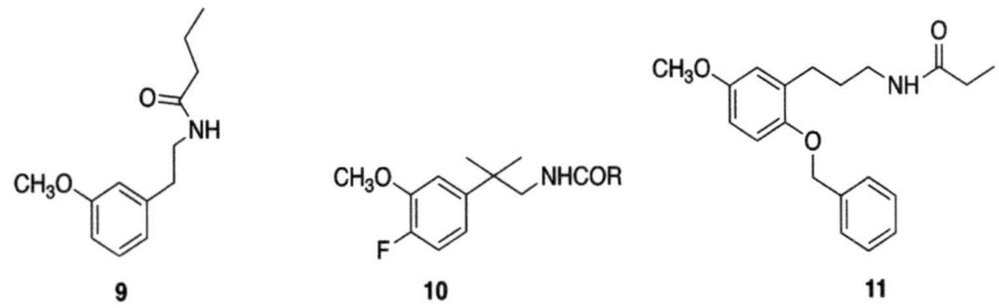

Figure 3.
Phenylalkyl amides.

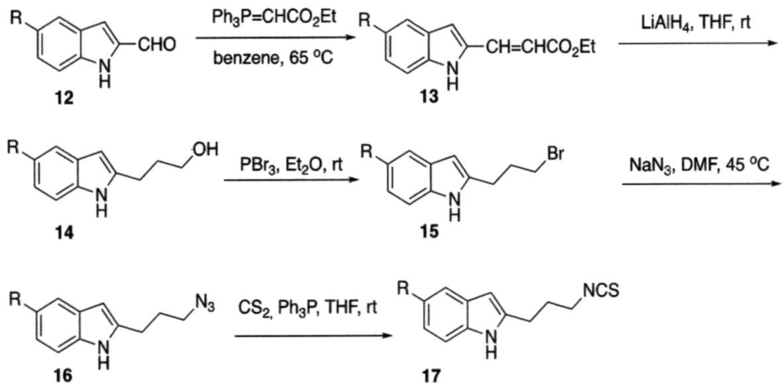

Figure 4.
C2-functionalized melatonin analogues.

MT1 receptor [4]. The synthesis of luzindole, achieved through a Pictet-Spengler reaction and formation of the intermediate β-carboline **19**, was first patented by Dubocovich et al. [40]. In 2008, Tsotinis et al. reported a new method of luzindole synthesis, through the C-3 indole nitroolefin **22**, leading to a much higher overall yield [41] (**Figure 5**).

The benzo[*b*]furan nucleus can replace the indole skeleton and retain its reactivity. 5-Methoxy-3-oxo-2,3-dihydrobenzo[*b*]furan (**25**) was prepared from 4-methoxyphenol (**23**) by acylation with chloracetonitrile followed by cyclization [42] (**Figure 6**).

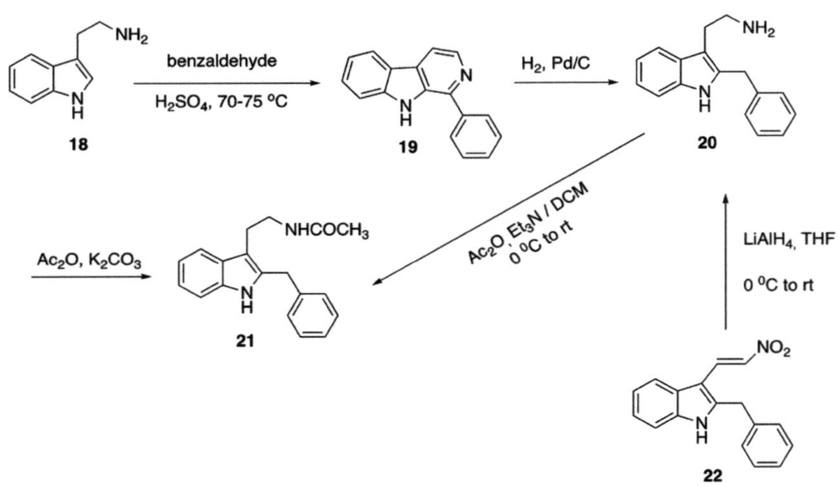

Figure 5.
Luzindole.

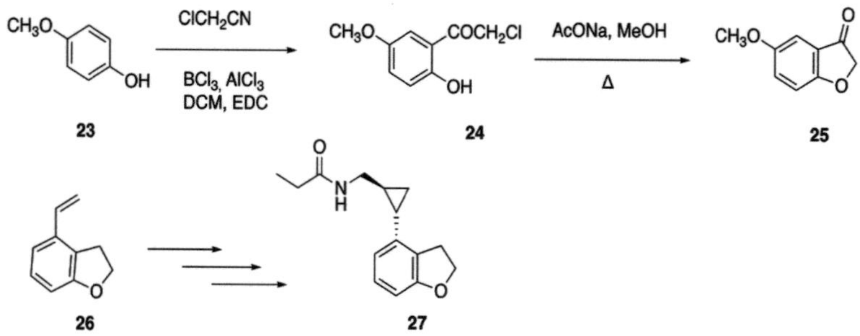

Figure 6.
Tasimelteon.

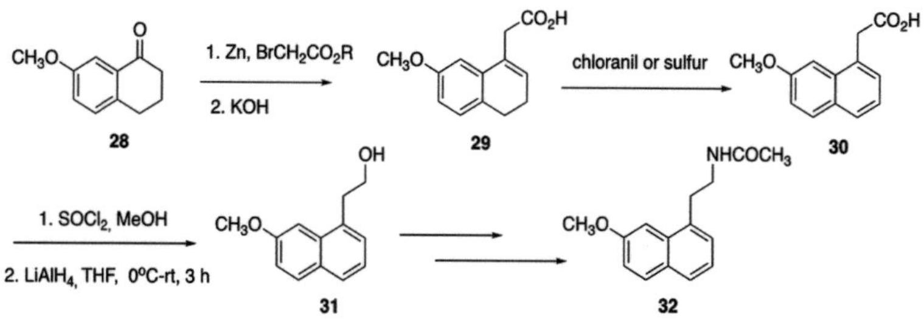

Figure 7.
Agomelatine.

Tasimelteon, *N*-[[(1*R*,2*R*)-2-(2,3-dihydro-1-benzofuran-4-yl)cyclopropyl] methyl] propenamide (**27**), is a melatonin agonist, which bears the benzo[*b*]furan skeleton and was approved by the FDA, in January 2014, for the treatment of non-24 h sleep–wake disorder [43]. The starting material for the synthesis of tasimelteon is the 4-vinyl-2, 3-dihydrobenzofuran (**26**).

The naphthalene scaffold can also be considered as a melatonin-acting biomolecule with high affinity and potency [44, 45]. The preparation of the key intermediate in this synthesis, 2-(7-methoxy-1-naphthyl)ethanol (**31**), is depicted in **Figure** 7. Agomelatine, *N*-[2-(7-methoxy-1-naphthyl)ethyl]acetamide (**32**), was recently approved for medical use in Europe and Australia [46].

3. Constrained polycyclic derivatives

Tricyclic and even larger constrained derivatives have been investigated for their melatoninergic potency. The synthesis of 6,7,8,9-tetrahydropyridino[1,2-*a*]indole (**36**) [47] is illustrated in **Figure 8**.

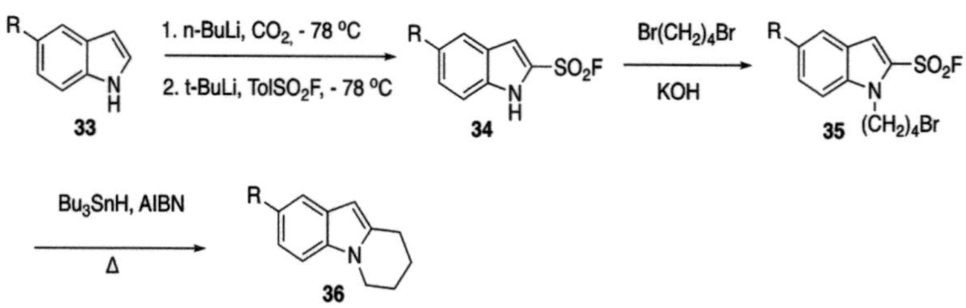

Figure 8.
6,7,8,9-Tetrahydropyridino[1,2-a]indole.

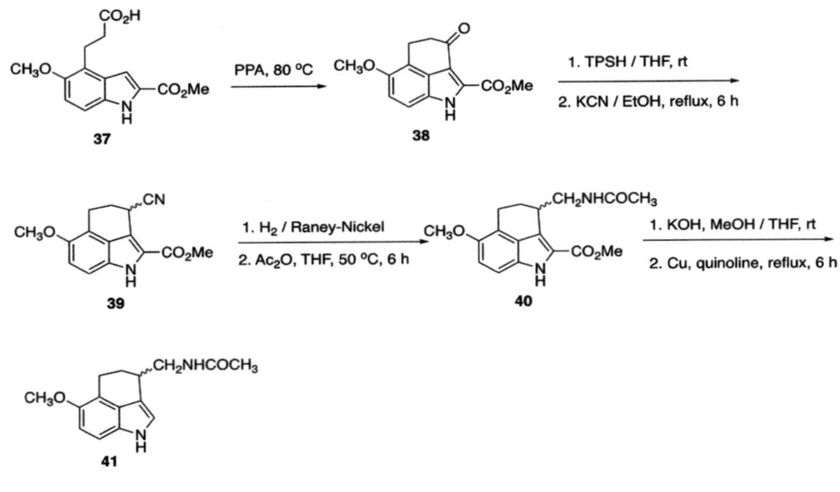

Figure 9.
1,3,4,5-Tetrahydro[cd]indole.

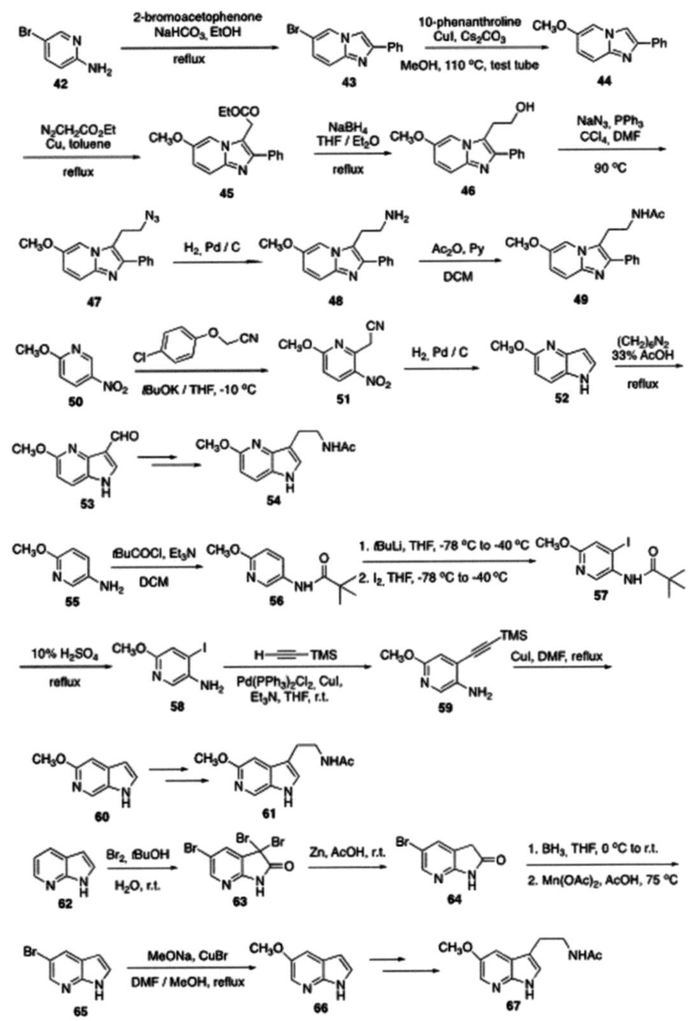

Figure 10.
Azaindoles.

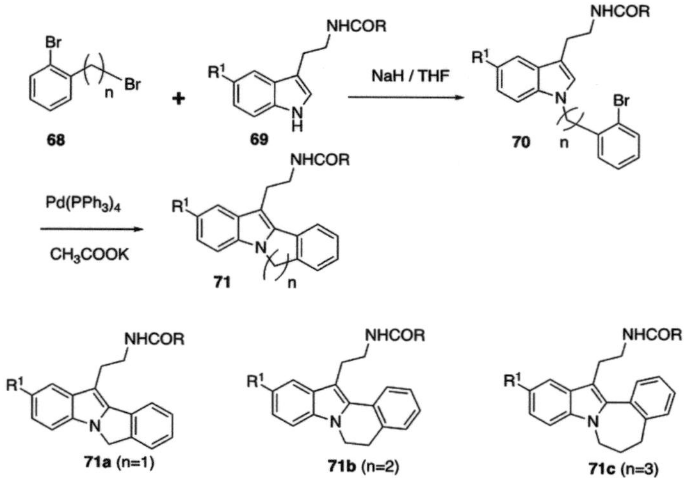

Figure 11.
Isoindolo[2,1-a]indoles and benzo[c]azepeno[2,1-a]indoles.

The 3-substituted 1,3,4,5-tetrahydro[*cd*]indoles exhibit higher melatonin receptor affinity than their more constrained congeners [30]. The key intermediate ketone **38** was obtained upon cyclization of the carboxylic acid **37** with polyphosphoric acid. As shown in **Figure 9**, the ketone **38** was converted to the corresponding cyanide **39**, in two steps. The latter gave then the respective acetamide **40**, and the final tricyclic adduct **41** was prepared by ester hydrolysis followed by decarboxylation of the corresponding acid in boiling quinoline in the presence of copper powder.

Azaindoles have also been proven to exhibit melatoninergic potency. Some melatonin analogues based on 3*a*-aza-, 4-aza-, 6-aza-, and 7-azaindole cores are described in **Figure 10**.

In the synthetic route to the 3*a*-azamelatonin analogue **49**, El Kazzouli et al. [48] reported the treatment of 2-amino-5-bromopyridine (**42**) with 2-bromoacetone and the use of ethyl 2-azidoacetate for the formation of the key intermediate ester **45**. In the synthesis of 3-substituted-4-azaindole **49**, Mazeas et al. [49] have used 2-methoxy-5-nitropyridine (**50**), as starting material, and standard chemistry procedures. The 4-azaindole analogue **50** was proven to be a stronger agonist than melatonin at both melatonin receptors [50]. The preparation of 6-azamelatonin derivative **61** involves the Sonogashira reaction, as reported in the literature [49]. Finally, the 7-azamelatonin congener **67** presents promising melatoninergic potential [49].

The isoindolo[2,1-*a*]indoles and benzo[*c*]azepeno[2,1-*a*]indoles were prepared by Tsotinis et al. [51]. The appropriate *N*-acetyl tryptamine was coupled with the respective dibromide **68**, and the derived *N*-alkyl indole **70** was then cyclized in the presence of $Pd(PPh_3)_4$ to afford the desired products **71 (Figure 11)**.

The pharmacological evaluation has shown that 6*H*-isoindolo[2,1-*a*]indoles (**71a**) are agonists, while the 5,6-dihydroindolo[2,1-*a*]isoquinolines (**71b**) are partial agonists/antagonists, and the 6,7-dihydro-5*H*-benzo[*c*]azepino[2,1-*a*]indoles (**71c**) are antagonists. Thus, the size of the linker between the phenyl ring and the pyrrole nitrogen atom serves as a switch pharmacological probe, spanning from agonist to antagonist melatoninergic action.

4. Chiral melatonin analogues

Some derivatives with constrained conformation also present chirality. Ramelteon is the most emblematic representative example of this class of

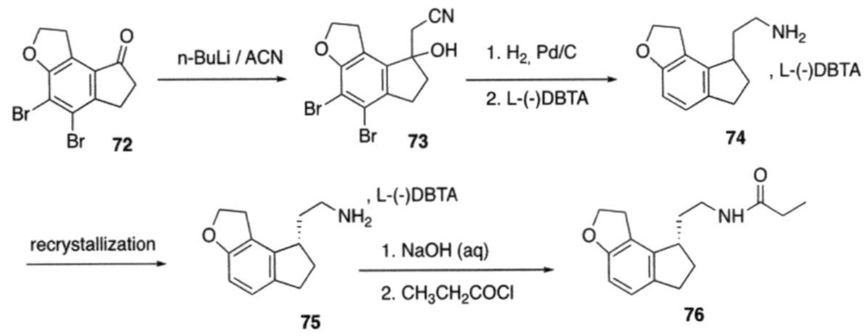

Figure 12.
Ramelteon.

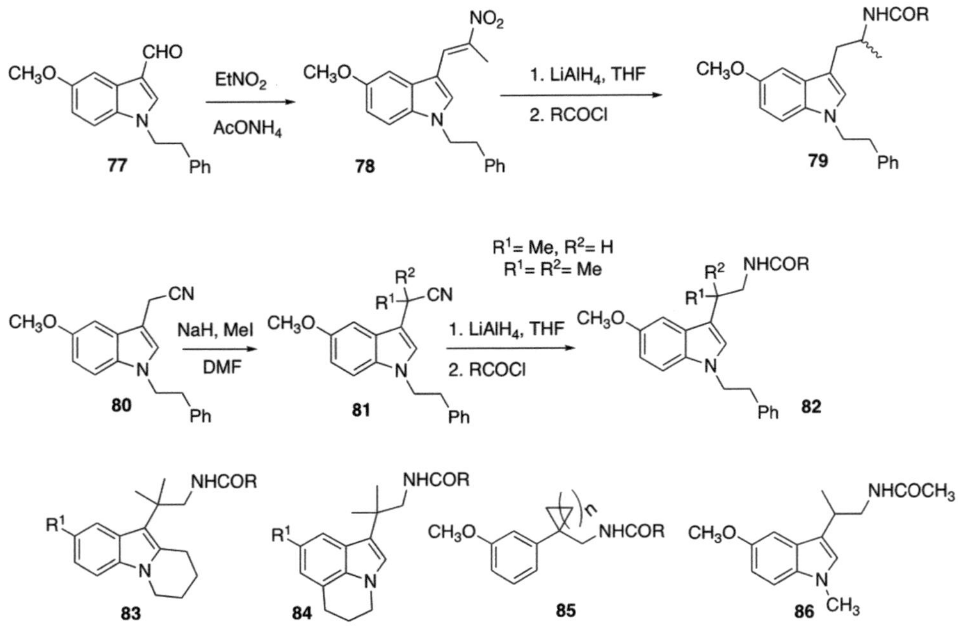

Figure 13.
Examples of chiral melatoninergic analogues and side chain conformationally constrained tricyclic derivatives 83 and 84.

compounds. Ramelteon, *N*-{2-[(8*S*)-1,6,7,8-tetrahydro-2H-indeno[5,4-*b*]furan-8-yl]ethyl}propanamide (**76**) [52], is a melatonin analogue approved by the FDA as a sedative-hypnotic. The following synthetic route [53], illustrated in **Figure 12**, uses dibenzoyl-*L*-tartaric acid as an acid to form the salt at the end of hydrogenation and as the resolution agent as well.

Most of these chiral derivatives are prepared as racemates and, then, in some cases, resolved. The racemate mixture of enantiomers provides an initial estimation of the biology of these compounds, although asymmetric syntheses may then be required if one of the enantiomers exhibits a selective result. Substituents on the 3-side chain, particularly at the β-position, present a preference for the active conformation. This hypothesis has been investigated by assessing the melatoninergic potency of various compounds which bear in their side chain small to large substituents. An example of α- and β-methyl side chain functionalized molecules with enhanced activity is the *N*1-phenethyl-substituted indole derivatives **79** and **82** [54]. The characteristic steps of the synthesis of these probes are illustrated in **Figure 13**. Similar results, in terms of activities and related conformation, have been obtained for the analogues **83, 84**, and **85** [55–57].

The *β*-methyl, *N*-methyl-substituted melatonin derivative **86** was prepared and resolved by chiral HPLC [58]. The (+) enantiomer has a tenfold higher potency in pigment aggregation in the *Xenopus laevis* protocol, while the (−) enantiomer has a 28-fold selectivity for the MT2 receptor.

5. Conclusions

A selection of key melatoninergic derivatives was reported herein. We pointed out the synthetic routes towards these melatonin analogues, first of the aromatic nucleus, then of the functionalities that have been introduced to the nucleus, and finally of those analogues with restrained conformations and those that are optically active. Much more needs to be explored about the variant functions of melatonin and through which receptor type they exert their action. The range of small molecules having agonist or antagonist effects on the melatonin receptors is large, and new scaffolds keep appearing as drug candidates in different treatments. This work is hoped to assist those seeking to explore the melatonin and melatoninergic field.

Conflict of interest

The authors declare no conflict of interest.

Author details

Andrew Tsotinis* and Ioannis P. Papanastasiou
Division of Pharmaceutical Chemistry, Department of Pharmacy, School of Health Sciences, National and Kapodistrian University of Athens, Athens, Greece

*Address all correspondence to: tsotinis@pharm.uoa.gr

References

[1] Zlotos DP, Jockers R, Cecon E, Rivara S, Witt-Enderby PA. MT 1 and MT 2 melatonin receptors: Ligands, models, oligomers, and therapeutic potential. Journal of Medicinal Chemistry. 2014;**57**(8):3161-3185. Available from: https://pubs.acs.org/doi/10.1021/jm401343c

[2] Oishi A, Cecon E, Jockers R. Melatonin receptor signaling: Impact of receptor oligomerization on receptor function. In: International Review of Cell and Molecular Biology. Cambridge, MA, United States: Elsevier; 2018. pp. 59-77. Available from: https://linkinghub.elsevier.com/retrieve/pii/S1937644818300121

[3] Gobbi G, Comai S. Sleep well. Untangling the role of melatonin MT1 and MT2 receptors in sleep. Journal of Pineal Research. 2019;**66**(3):e12544. Available from: http://doi.wiley.com/10.1111/jpi.12544

[4] Dubocovich ML, Delagrange P, Krause DN, Sugden D, Cardinali DP, Olcese J. International Union of Basic and Clinical Pharmacology. LXXV. Nomenclature, classification, and pharmacology of G protein-coupled melatonin receptors. Pharmacological Reviews. 2010;**62**(3):343-380. Available from: http://pharmrev.aspetjournals.org/lookup/doi/10.1124/pr.110.002832

[5] Stauch B, Johansson LC, McCorvy JD, Patel N, Han GW, Huang X-P, et al. Structural basis of ligand recognition at the human MT1 melatonin receptor. Nature. 2019;**569**(7755):284-288. Available from: http://www.nature.com/articles/s41586-019-1141-3

[6] Johansson LC, Stauch B, McCorvy JD, Han GW, Patel N, Huang X-P, et al. XFEL structures of the human MT2 melatonin receptor reveal the basis of subtype selectivity. Nature. 2019;**569**(7755):289-292. Available from: http://www.nature.com/articles/s41586-019-1144-0

[7] Tan D-X, Manchester LC, Terron MP, Flores LJ, Tamura H, Reiter RJ. Melatonin as a naturally occurring co-substrate of quinone reductase-2, the putative MT 3 melatonin membrane receptor: Hypothesis and significance. Journal of Pineal Research. 2007;**43**(4):317-320. Available from: http://doi.wiley.com/10.1111/j.1600-079X.2007.00513.x

[8] Ebisawa T, Karne S, Lerner MR, Reppert SM. Expression cloning of a high-affinity melatonin receptor from Xenopus dermal melanophores. Proceedings of the National Academy of Sciences. 1994;**91**(13):6133-6137. Available from: http://www.pnas.org/cgi/doi/10.1073/pnas.91.13.6133

[9] Reppert SM, Godson C, Mahle CD, Weaver DR, Slaugenhaupt SA, Gusella JF. Molecular characterization of a second melatonin receptor expressed in human retina and brain: The Mel1b melatonin receptor. Proceedings of the National Academy of Sciences. 1995;**92**(19):8734-8738. Available from: http://www.pnas.org/cgi/doi/10.1073/pnas.92.19.8734

[10] Reppert SM, Weaver DR, Ebisawa T. Cloning and characterization of a mammalian melatonin receptor that mediates reproductive and circadian responses. Neuron. 1994;**13**(5):1177-1185. Available from: https://linkinghub.elsevier.com/retrieve/pii/0896627394900558

[11] Vakkuri O, Leppäluoto J, Vuolteenaho O. Development and validation of a melatonin radioimmunoassay using radioiodinated melatonin as tracer. Acta Endocrinologica. 1984;**106**(2):152-157. Available from: https://eje.bioscientifica.com/view/journals/eje/106/2/acta_106_2_002.xml

[12] Vaněček J, Pavlík A, Illnerová H. Hypothalamic melatonin receptor sites revealed by autoradiography. Brain Research. 1987;**435**(1-2):359-362. Available from: https://linkinghub.elsevier.com/retrieve/pii/0006899387916258

[13] Mor M, Rivara S, Silva C, Bordi F, Plazzi PV, Spadoni G, et al. Melatonin receptor ligands: Synthesis of New melatonin derivatives and comprehensive comparative molecular field analysis (CoMFA) study. Journal of Medicinal Chemistry. 1998;**41**(20):3831-3844. Available from: https://pubs.acs.org/doi/10.1021/jm9810093

[14] Depreux P, Lesieur D, Mansour HA, Morgan P, Howell HE, Renard P, et al. Synthesis and structure-activity relationships of novel naphthalenic and bioisosteric related amidic derivatives as melatonin receptor ligands. Journal of Medicinal Chemistry. 1994;**37**(20):3231-3239. Available from: https://pubs.acs.org/doi/abs/10.1021/jm00046a006

[15] Langlois M, Bremont B, Shen S, Poncet A, Andrieux J, Sicsic S, et al. Design and synthesis of new naphthalenic derivatives as ligands for 2-[^{125}I] iodomelatonin binding sites. Journal of Medicinal Chemistry. 1995;**38**(12):2050-2060. Available from: https://pubs.acs.org/doi/abs/10.1021/jm00012a004

[16] Garratt PJ, Travard S, Vonhoff S, Tsotinis A, Sugden D. Mapping the melatonin receptor. 4. Comparison of the binding affinities of a series of substituted Phenylalkyl amides. Journal of Medicinal Chemistry. 1996;**39**(9):1797-1805. Available from: http://pubs.acs.org/doi/abs/10.1021/jm9508189

[17] Plieninger H. The chemistry of indoles. Organic chemistry, a series of monographs, Vol. 18. Von R. J. Sundberg. Academic Press, New York–London 1970. 1. Aufl., X, 489 S., s229/—. Angewandte Chemie. 1971;**83**(9):338-338. Available from: http://doi.wiley.com/10.1002/ange.19710830918

[18] Gribble GW. Recent developments in indole ring synthesis—Methodology and applications. Contemporary Organic Synthesis. 1994;**1**(3):145-172. Available from: http://xlink.rsc.org/?DOI=CO9940100145

[19] Gribble GW. Recent developments in indole ring synthesis—Methodology and applications. Journal of the Chemical Society, Perkin Transactions 1. 2000;**1**(7):1045-1075. Available from: http://xlink.rsc.org/?DOI=a909834h

[20] Frohn MA, Seaborn CJ, Johnson DW, Phillipou G, Seamark RF, Matthews CD. Structure - activity relationship of melatonin analogues. Life Sciences. 1980;**27**(22):2043-2046. Available from: https://linkinghub.elsevier.com/retrieve/pii/0024320580904828

[21] Heward CB, Hadley ME. Structure-activity relationships of melatonin and related indoleamines. Life Sciences. 1975;**17**(7):1167-1177. Available from: https://linkinghub.elsevier.com/retrieve/pii/0024320575903409

[22] Leclerc V, Fourmaintraux E, Depreux P, Lesieur D, Morgan P, Howell HE, et al. Synthesis and structure–activity relationships of novel naphthalenic and bioisosteric related amidic derivatives as melatonin receptor ligands. Bioorganic & Medicinal Chemistry. 1998;**6**(10):1875-1887. Available from: https://linkinghub.elsevier.com/retrieve/pii/S0968089698001473

[23] Fukatsu K, Uchikawa O, Kawada M, Yamano T, Yamashita M, Kato K, et al. Synthesis of a novel series of benzocycloalkene derivatives as melatonin receptor agonists. Journal of Medicinal Chemistry. 2002;**45**(19):4212-4221. Available from: https://pubs.acs.org/doi/10.1021/jm020114g

[24] Spadoni G, Mor M, Tarzia G. Structure-affinity relationships of indole-based melatonin analogs. Neurosignals. 1999;**8**(1-2):15-23. Available from: https://www.karger.com/Article/FullText/14564

[25] Garratt PJ, Jones R, Rowe SJ, Sugden D. Mapping the melatonin receptor. 1. The 5-methoxyl group of melatonin is not an essential requirement for biological activity. Bioorganic & Medicinal Chemistry Letters. 1994;**4**(13):1555-1558. Available from: https://linkinghub.elsevier.com/retrieve/pii/S0960894X01805655

[26] Garratt PJ, Vonhoff S, Rowe SJ, Sugden D. Mapping the melatonin receptor. 2. Synthesis and biological activity of indole derived melatonin analogues with restricted conformations of the C-3 amidoethane side chain. Bioorganic & Medicinal Chemistry Letters. 1994;**4**(13):1559-1564. Available from: https://linkinghub.elsevier.com/retrieve/pii/S0960894X01805667

[27] Garratt PJ, Jones R, Tocher DA, Sugden D. Mapping the melatonin receptor. 3. Design and synthesis of melatonin agonists and antagonists derived from 2-Phenyltryptamines. Journal of Medicinal Chemistry. 1995;**38**(7):1132-1139. Available from: https://pubs.acs.org/doi/abs/10.1021/jm00007a010

[28] Sugden D, Rowe SJ. 2-Iodo-*N*-butanoyl-5-methoxytryptamine: A potent melatonin receptor agonist. Pharmacology Communications. 1994;**4**:267-276

[29] Mathé-Allainmat M, Gaudy F, Sicsic S, Dangy-Caye A-L, Shen S, Brémont B, et al. Synthesis of 2-amido-2,3-dihydro-1*H*-phenalene derivatives as New Conformationally restricted ligands for melatonin receptors. Journal of Medicinal Chemistry. 1996;**39**(16):3089-3095. Available from: https://pubs.acs.org/doi/10.1021/jm960219h

[30] Spadoni G, Balsamini C, Diamantini G, Di Giacomo B, Tarzia G, Mor M, et al. Conformationally restrained melatonin analogues: Synthesis, binding affinity for the melatonin receptor, evaluation of the biological activity, and molecular Modeling study. Journal of Medicinal Chemistry. 1997;**40**(13):1990-2002. Available from: https://pubs.acs.org/doi/10.1021/jm960651z

[31] Flaugh ME, Crowell TA, Clemens JA, Sawyer BD. Synthesis and evaluation of the antiovulatory activity of a variety of melatonin analogs. Journal of Medicinal Chemistry. 1979;**22**(1):63-69. Available from: https://pubs.acs.org/doi/abs/10.1021/jm00187a015

[32] Sugden D, Chong NWS, DFV L. Structural requirements at the melatonin receptor. British Journal of Pharmacology. 1995;**114**(3):618-623. Available from: http://doi.wiley.com/10.1111/j.1476-5381.1995.tb17184.x

[33] Li JJ. Batcho–Leimgruber indole synthesis. In: Name Reactions. Cham: Springer International Publishing; 2014. pp. 34-35. Available from: http://link.springer.com/10.1007/978-3-319-03979-4_17

[34] Repke DB, Ferguson WJ. Psilocin analogs. III. Synthesis of 5-methoxy- and 5-hydroxy-1,2,3,4-tetrahydro-9*H*-pyrido[3,4-*b*]indoles. Journal of Heterocyclic Chemistry. 1982;**19**(4):845-848. Available from: http://doi.wiley.com/10.1002/jhet.5570190428

[35] Ates-Alagoz Z, Buyukbingol Z, Buyukbingol E. Synthesis and antioxidant properties of some indole ethylamine derivatives as melatonin analogs. Die Pharmazie. 2005 Sep;**60**(9):643-647

[36] Righi M, Topi F, Bartolucci S, Bedini A, Piersanti G, Spadoni G. Synthesis of tryptamine derivatives via a direct, one-pot reductive alkylation

of indoles. The Journal of Organic Chemistry. 2012;**77**(14):6351-6357. Available from: https://pubs.acs.org/doi/10.1021/jo3010028

[37] Hu Y, Ho MKC, Chan KH, New DC, Wong YH. Synthesis of substituted *N*-[3-(3-methoxyphenyl)propyl] amides as highly potent MT2-selective melatonin ligands. Bioorganic & Medicinal Chemistry Letters. 2010;**20**(8):2582-2585. Available from: https://linkinghub.elsevier.com/retrieve/pii/S0960894X10002842

[38] Tsotinis A, Kompogennitaki R, Papanastasiou I, Garratt PJ, Bocianowska A, Sugden D. Fluorine substituted methoxyphenylalkyl amides as potent melatonin receptor agonists. Medicinal Chemistry Communications. 2019;**10**(3):460-464. Available from: https://pubs.rsc.org/en/content/articlelanding/2019/md/c8md00604k

[39] Tsotinis A, Afroudakis PA, Davidson K, Prashar A, Sugden D. Design, synthesis, and melatoninergic activity of New Azido- and isothiocyanato-substituted indoles. Journal of Medicinal Chemistry. 2007;**50**(25):6436-6440. Available from: https://pubs.acs.org/doi/10.1021/jm7010723

[40] Dubocovich ML, Rajadhyaksha VJ, Helmy AA. 2-Aryl *N*-acetyltryptamines and Process of Preparing Such. US 528334; 1994

[41] Tsotinis A, Afroudakis P. Melatonin receptor antagonist luzindole: A facile new synthesis. Letters in Organic Chemistry. 2008;**5**(6):507-509. Available from: http://www.eurekaselect.com/openurl/content.php?genre=article&issn=1570-1786&volume=5&issue=6&spage=507

[42] Hammond ML, Zambias RA, Chang MN, Jensen NP, McDonald J, Thompson K, et al. Antioxidant-based inhibitors of leukotriene biosynthesis. The discovery of 6-[1-[2-(hydroxymethyl) phenyl]-1-propen-3-yl]-2,3-dihydro-5-benzofuranol, a potent topical antiinflammatory agent. Journal of Medicinal Chemistry. 1990;**33**(3):908-918. Available from: https://pubs.acs.org/doi/abs/10.1021/jm00165a005

[43] Hardeland R. Tasimelteon, a melatonin agonist for the treatment of insomnia and circadian rhythm sleep disorders. Current Opinion in Investigational Drugs. 2009;**10**(7):691-701

[44] Yous S, Andrieux J, Howell HE, Morgan PJ, Renard P, Pfeiffer B, et al. Novel naphthalenic ligands with high affinity for the melatonin receptor. Journal of Medicinal Chemistry. 1992;**35**(8):1484-1486. Available from: https://pubs.acs.org/doi/abs/10.1021/jm00086a018

[45] Li P-K, Chu G-H, Gillen ML, Parekh T, Witt-Enderby PA. The development of a charged melatonin receptor ligand. Bioorganic & Medicinal Chemistry Letters. 1997;**7**(18):2409-2414. Available from: https://linkinghub.elsevier.com/retrieve/pii/S0960894X97004447

[46] Norman TR, Olver JS. Agomelatine for depression: Expanding the horizons? Expert Opinion on Pharmacotherapy. 2019;**20**(6):647-656. Available from: https://www.tandfonline.com/doi/full/10.1080/14656566.2019.1574747

[47] Tsotinis A, Panoussopoulou M, Sivananthan S, Sugden D. Synthesis of new tricyclic melatoninergic ligands. Il Farmaco. 2001;**56**(9):725-729. Available from: https://linkinghub.elsevier.com/retrieve/pii/S0014827X01011259

[48] El Kazzouli S, Griffon du Bellay A, Berteina-Raboin S, Delagrange P, Caignard D-H, Guillaumet G. Design and synthesis of 2-phenylimidazo[1,2-*a*] pyridines as a novel class of melatonin

receptor ligands. European Journal of Medicinal Chemistry. 2011;**46**(9): 4252-4257. Available from: https://linkinghub.elsevier.com/retrieve/pii/S022352341100506X

[49] Viaud-Massuard M-C, Mazéas D, Guillaumet G. Synthesis of new melatoninergic ligands including azaindole moiety. Heterocycles. 1999;**50**(2):1065. Available from: http://www.heterocycles.jp/newlibrary/libraries/abst/07243

[50] Jeanty M, Suzenet F, Delagrange P, Nosjean O, Boutin JA, Caignard DH, et al. Design and synthesis of 1-(2-alkanamidoethyl)-6-methoxy-7-azaindole derivatives as potent melatonin agonists. Bioorganic & Medicinal Chemistry Letters. 2011;**21**(8):2316-2319. Available from: https://linkinghub.elsevier.com/retrieve/pii/S0960894X11002812

[51] Faust R, Garratt PJ, Jones R, Yeh L-K, Tsotinis A, Panoussopoulou M, et al. Mapping the melatonin receptor. 6. Melatonin agonists and antagonists derived from 6*H*-isoindolo[2,1-*a*]indoles, 5,6-dihydroindolo[2,1-*a*]isoquinolines, and 6,7-dihydro-5*H*-benzo[c]azepino[2,1-*a*]indoles. Journal of Medicinal Chemistry. 2000;**43**(6):1050-1061

[52] Uchikawa O, Fukatsu K, Tokunoh R, Kawada M, Matsumoto K, Imai Y, et al. Synthesis of a novel series of tricyclic indan derivatives as melatonin receptor agonists. Journal of Medicinal Chemistry. 2002;**45**(19):4222-4239. Available from: https://pubs.acs.org/doi/10.1021/jm0201159

[53] Xiao S, Chen C, Li H, Lin K, Zhou W. A novel and practical synthesis of Ramelteon. Organic Process Research & Development. 2015;**19**(2):373-377. Available from: https://pubs.acs.org/doi/10.1021/op500386g

[54] Sugden D, Davies DJ, Garratt PJ, Jones R, Vonhoff S. Radioligand binding affinity and biological activity of the enantiomers of a chiral melatonin analogue. European Journal of Pharmacology. 1995;**287**(3): 239-243. Available from: https://linkinghub.elsevier.com/retrieve/pii/0014299995004890

[55] Tsotinis A, Panoussopoulou M, Hough K, Sugden D. Synthesis and biological evaluation of new β,β'-disubstituted 6,7,8,9-tetrahydropyrido[1,2-*a*]indol-10-yl ethylamido melatoninergic ligands. European Journal of Pharmaceutical Sciences. 2003;**18**(5):297-304. Available from: https://linkinghub.elsevier.com/retrieve/pii/S0928098703000204

[56] Tsotinis A, Panoussopoulou M, Eleutheriades A, Davidson K, Sugden D. Design, synthesis and melatoninergic activity of new unsubstituted and β,β'-difunctionalised 2,3-dihydro-1*H*-pyrrolo[3,2,1-*ij*]quinolin-6-alkanamides. European Journal of Medicinal Chemistry. 2007;**42**(7):1004-1013. Available from: https://linkinghub.elsevier.com/retrieve/pii/S0223523407000372

[57] Tsotinis A, Vlachou M, Papahatjis D, Nikas S, Sugden D. An efficient synthesis of simple β,β-Cyclobisalkylated Melatoninergic Phenylalkylamides. Letters in Organic Chemistry 2007;**4**(2):92-95. Available from: http://www.eurekaselect.com/openurl/content.php?genre=article&issn=1570-1786&volume=4&issue=2&spage=92

[58] Tsotinis A, Vlachou M, Papahatjis DP, Calogeropoulou T, Nikas SP, Garratt PJ, et al. Mapping the melatonin receptor. 7. Subtype selective ligands based on β-substituted *N*-Acyl-5-methoxytryptamines and β-substituted *N*-Acyl-5-methoxy-1-methyltryptamines. Journal of Medicinal Chemistry. 2006;**49**(12):3509-3519. DOI: 10.1021/jm0512544

Section 2

Insomnia Related Dysfunctions

Chapter 2

Clinical Use of Melatonin in the Treatment of Sleep Disorders

Alexander Zakharov and Elena Khivintseva

Abstract

Sleep disorders are a group of conditions that affect the circadian rhythm of sleep-wake, leading to social and professional maladaptation. At the moment, there is a wide range of medications aimed at the treatment of sleep disorders, but the results from their use are not always satisfactory. Benzodiazepines, antidepressants, and antihistamines may cause dependence or withdrawal effects. Melatonin (N-acetyl-5-methoxytryptamine) is an endogenous hormone produced by the pineal gland that affects intraday, seasonal rhythm, and the sleep-wake cycle. Studies of the effects of melatonin have demonstrated its ability to synchronize circadian rhythms, reduce the latency of slow sleep, increase the duration of sleep, and improve its subjective quality. This review highlights the current therapeutic possibilities of using melatonin in various sleep disorders, taking into account the mechanisms of its action. Also, the prospects of using melatonin due to its chronobiological effect in other sleep disorders, such as parasomnia, sleep-dependent respiratory disorders, and hypersomnia, are emphasized. At the moment, melatonin is one of the methods for correcting intraday rhythms and some types of insomnia.

Keywords: sleep, melatonin, sleep disorders, sleep-wake cycle

1. Introduction

Sleep is fundamental to the mental and physical health of a person. Lack of sleep is a significant risk factor for obesity, diabetes, diseases of the cardiovascular system, as well as anxiety and depressive disorders. Sleep disorders have a significant financial burden on the healthcare system and complicate the treatment of major somatic diseases. Sleep disorders are a category of diseases that include hypersomnia, insomnia (accompanied by difficulty falling asleep, maintaining sleep, and early awakening), circadian rhythm disturbance, parasomnia, and sleep-dependent breathing disorders. The consequence of some sleep disorders is a violation of falling asleep and maintaining sleep, drowsiness, and, as a consequence, a decrease in the quality of life. Some sleep disorders can also lead to severe impaired ability to perform every day and professional tasks related to concentration, switching attention, and spatial perception [1].

The development of pharmacological treatment methods has provoked an increase in the frequency of sleep disorders in the last decade, as a result of undesirable effects of this therapy. The most common disease is insomnia, which according to the classification criteria for mental disorders *Diagnostic and Statistical Manual of Mental Disorders IV (DSM-IV)* in the general population is found in 4–6%.

The main classes of drugs for the treatment of insomnia are barbiturates, benzodiazepines, benzodiazepine agonists, antidepressants, and anxiolytics. These drugs can cause a large number of side effects associated with excessive daytime sleepiness, decreased concentration, and switching attention and can cause deterioration of short-term memory. In some cases, with prolonged use of these drugs, dependence may form, and with cancelation, a "rebound phenomenon" may occur. In this regard, it becomes relevant to search for new pharmaceuticals that reduce the number and severity of these side effects while maintaining the proper level of effectiveness. One of these drugs, with long-term administration of minimal side effects and sufficient effectiveness in certain sleep disorders, is melatonin. Melatonin is mainly produced by the pineal gland with a peak of activity at night; the concentration fluctuation coincides with the circadian rhythm. Melatonin-based preparations have good tolerance in various age periods, without forming dependency [2–4].

Other effects are inherent to melatonin, namely, regulation of circadian, seasonal rhythms; regulation of the psychoemotional and cognitive sphere; antioxidant, neuroprotective, and geroprotective effect; immunomodulatory; vegetative stabilizing; and oncological and stress-protective effect.

The multiplicity of effects of melatonin is due to the large number of targets on which this hormone has an effect. The most studied mechanism for the implementation of the action of melatonin remains its effect on suprachiasmal nuclei (SCN) of the hypothalamus. Through SCN, the chronobiological effect of melatonin is realized and, of course, its hypnotic effects. Melatonin interacts with two types of G-protein-bound receptors—MT1 and MT2 [5]. MT1-type receptors are distributed in the hippocampus, caudate nucleus, pillow, suprachiasmatic nuclei, paraventricular nucleus, supraoptic nucleus, Meynert nucleus, adjacent nucleus, substantia nigra, mammary bodies, and retina. MT2-type receptors are mainly detected in the hippocampus, SCN, and the retina. Both types of receptors are expressed by neurons and glial cells of the cerebral and cerebellar cortex, in the thalamus, and pineal gland [5, 6].

Melatonin is released into the blood plasma as a rhythmic oscillatory pattern, which is regulated by SCN neurons. Daylight suppresses the release of melatonin through the retinohypothalamic tract, projecting from melanopsin-expressing retinal ganglion cells to SCN neurons. It is known, for example, that night illumination is 2000–2500 Lux within 2 hours, which completely inhibits the secretion of melatonin. On the other hand, traditional home light (50–300 Lux) practically does not have a suppressive effect on the secretion of melatonin [7]. The neural relationship between the structures of the central nervous system, where axons of melanopsin-expressing ganglion cells are projected, primarily with SCN neurons and the sympathetic nervous system, is via the superior cervical sympathetic ganglion, from where the nerve fibers go directly to the pinealocytes and regulate the exocytosis of norepinephrine, which activates melatonin synthesis and its release [8]. As mentioned above, melatonin easily penetrates through biological barrier: it is secreted continuously into the blood plasma and enters various fluids (saliva, urine, cerebrospinal fluid, preovulatory follicle, spermatozoa, amniotic fluid, and human milk). The maximum level of melatonin in blood plasma is at 03.00–04.00 at night. The indicator varies depending on the chronotype and is not determined in the daytime. Melatonin levels have a pronounced intersubject heterogeneity but are steadily repeated in the same person. After birth, the rhythmic production of melatonin during the day reaches very high levels by 3–6 years of life and then decreases by almost 80% to levels in an adult. The melatonin rhythm is generated by the endogenous clock of the hypothalamic SCN neurons, which are affected by the light/dark cycle (zeitgeber). Seasonal effects on the secretion of melatonin are manifested in an increase in nighttime secretion of melatonin, which is associated with a decrease in plasma of ovarian steroids. On the other hand, urban lighting reduces seasonal differences in the secretion of melatonin, cortisol, and thyrotropin. Winter-type seasonal affective disorders are characterized by recurrent depressive episodes during a short photoperiod.

Melatonin, due to its amphotericity (amphiphilicity), is able to penetrate into the cell, organelles, and nuclear membranes and directly interacts with intracellular molecules, exerting a non–receptor-mediated effect. Along with this, melatonin exerts a receptor-mediated effect on target cells, as a result of the interaction of the hormone with either membrane or nuclear receptors [9]. The main physiological functions of melatonin are due to its hormonal properties; however, the hormone also has an autocrine and paracrine effect, in particular in the retina and gastrointestinal tract [10].

Outside of SCN, MT1 and MT2 receptors are also found in large numbers in the duodenum, colon, cecum and appendix, gallbladder epithelium, parotid gland, pancreas, β-cells of the endocrine system, pancreas, coronary, and cerebral arteries adipose tissue. In addition to membrane receptors for melatonin, there are also nuclear receptors: RORα and RORβ. The prevalence of RORα is highest in T and B lymphocytes, neutrophils, and monocytes, whereas RORβ are found mainly in the brain, pineal gland, retina, and spleen.

The modulating effect on sleep architecture is also realized by melatonin due to membrane receptors MT1 and MT2. The activation of the MT2 receptor contributes to increasing the duration of slow-wave sleep. The activation of the MT 1 receptor has a decrease in the duration of slow-wave sleep [11, 12].

The effects of melatonin, in addition to effects on SCN, on neural networks of passive brain function default mode network (DMN) were also demonstrated. Their activation is accompanied by the appearance of a feeling of fatigue and is characterized by changes typical of sleep in such parts of the cortex as the precuneus located in the rostromedial aspect of the occipital cortex [13, 14]. Because the general effect of melatonin through two membrane receptors does not increase the duration of slow-wave sleep (SWS) [15], the main effect of melatonin is not associated with its homeostatic effect on sleep. Therefore, its effect can be attributed to sleep regulation through the circadian component [16].

The multiple representation of melatonin receptors in the central nervous system, its effect on one of the key components of the regulation of the sleep-wake cycle, leads to the multiplicity of the clinical use of this hormone, especially in pathological conditions accompanied by primary or secondary circadian rhythm disturbances.

2. Melatonin and sleep disorders

2.1 Melatonin and disorders of the sleep-wake cycle

Circadian disturbances of the sleep-wake rhythm are associated with disconnection of the synchronization of the endogenous circadian rhythm and environmental influences. Melatonin signals the onset of darkness, and activation of its production indirectly depends on the activity of intrinsically photosensitive retinal ganglion cells (ipRGC) or true light-sensitive retinal ganglion cells. However, there is also an endogenous melatonin release profile that allows SCN activation regardless of external light, maintaining sleep-wake rhythms and neuroendocrine rhythms in a 24-hour cycle. However, the absence of external-stabilizing effect of zeitgeber (daily light change) can lead to the formation of a non-24-hour sleep-wake cycle. For example, in completely blind subjects, it is quite common (in 50–75% of cases) to observe a non-24-hour sleep-wake disorder (non-24-hour sleep-wake disorder), the occurrence of which is associated with the inability to synchronize with changes in light [17]. Circadian rhythm disorders can be divided into conditions that may be caused by endogenous or exogenous factors. The first subgroup includes the syndrome of delayed onset of sleep and wake phases (advanced sleep-wake phase disorder), early onset of the sleep phase (delayed sleep-wake phase disorder), irregular sleep-wake rhythm disorder (irregular sleep-wake rhythm disorder), and non-24-hour sleep-wake cycle (non-24-hour sleep-wake disorder). The group with exogenous causes of occurrence includes jet lag disorder, a disorder caused by a shift work schedule (shift work disorder) or a result of behavioral features of going to bed and violation of the work and rest regime in the format of about 24-hour circadian rhythm. The circadian rhythm is regulated by melatonin, while the production of melatonin itself is regulated by external influences, the most important of which is the effect of light, which activates the retinal ganglion cells containing the light-sensitive pigment melanopsin. External influences with excessive activation of signal systems implemented through SCN excitation are caused by the lifestyle of modern people, the use of electronic devices. Such excessive activation can lead to difficulty in initiating sleep, reducing its duration [18]. A decrease in melatonin secretion serves as one of the main mechanisms for the occurrence of such a disorder as delayed sleep phases [19]. There is a positive modulating effect of melatonin on the circadian rhythm of sleep-wakefulness and sleep efficiency both in pathology and in healthy subjects [20].

In separate studies in patients with delayed onset of sleep and wake phases in combination with attention deficit hyperactivity disorder, therapy was performed at a dose of 10 mg, lasting for 4 or more years. The therapeutic effect of melatonin was shown in reducing the start time of sleep and increasing the time of wakefulness in these patients. The use of melatonin in a dose of 3 mg for the treatment of disorders of the sleep-wake cycle in children did not show any effects on the process of puberty in the long-term period. However, it should be noted that these studies are isolated and do not carry a sufficiently high level of evidence [21]. However, even this long-term use of melatonin was not accompanied by any significant or serious adverse events.

Table 1 presents data on the efficacy of melatonin and its agonists in various forms of sleep-wake disorder [21].

According to the recommendations for the treatment of these conditions, melatonin and its agonists have a sufficient level of evidence when applied to the diagnosis of delayed onset of sleep and wake phases and irregular sleep-wake

Type of disorder (syndrome)	Efficiency	Level of evidence
Delayed onset of the phases of sleep and wakefulness	Recommended for adults with or without depression Recommended for children and adolescents without or with concomitant psychiatric pathology	Low
	Recommended for children and adolescents without or with concomitant psychiatric pathology	Moderate
Non-24-hour sleep-wake cycle	Recommended for blind adults	Low
Irregular rhythm of sleep-wakefulness	Not recommended for seniors with dementia. Recommended for children and adolescents with neurological pathology	Moderate

Table 1.
The use of melatonin and its agonists in various types of disorders of the sleep-wake cycle.

rhythm syndrome. Concerning the recommendations on the dose of melatonin, no consensus has been formed, since in studies on the basis of which recommendations are formed with the use of a wide variety of doses of melatonin, from 0.3 to 10 mg. For the non-24-hour sleep-wake cycle syndrome alone, in 2014, the US Food and Drug Administration (FDA) approved a melatonin agonist (tasimelteon) as a therapy.

However, individual studies have demonstrated a high therapeutic effect in the treatment of completely blind patients with N24HSWD immediate-release melatonin preparations. Taking a 0.5–10 mg of melatonin helped accelerate the synchronization of the endogenous sleep-wake rhythm with a 24-hour rhythm, according to the profile of the production of melatonin and cortisol. Also, separate studies demonstrate that drugs with modified melatonin release can also be effective in stabilizing circadian rhythms in completely blind patients with N24HSWD [22].

The so-called sleep-wake cycle disturbance states, namely, "jetlag," which occurs when changing time zones during an eastbound flight, can be corrected quite well with exogenous melatonin. In separate studies, various doses of melatonin (from 0.5 to 10 mg) used at bedtime, 3 days before the transmeridian flight and 5 days after it, were used to treat jetlag [23, 24]. The effectiveness of melatonin in most studies was already shown during the first 3 days after the completed flight, but subsequently, patients who did not take melatonin showed the same sleep-wake cycle characteristics as the group of people taking it. The main effect of melatonin in the first 3 days after the transmeridian flight was an increase in the duration and quality of night sleep, based both on subjective sensations and on the data of objective methods for recording sleep patterns (polysomnography and actigraphy) [25, 26].

At the same time, melatonin had a positive effect on latency and duration of sleep. Melatonin agonists have also shown their effectiveness in accelerating adaptation to a new time zone. Melatonin agonists (ramelteon and tasimelteon) are approved by the FDA for the treatment of time zone change syndrome ("jet lag"). As a pharmacological method of treating these types of disorders, the use of agomelatine, long-acting melatonin, and tasimelteon was approved by the European Medicines Agency. Most studies have evaluated the effects of melatonin on jet lag when changing time zones eastward, but there are also few studies showing its effectiveness in treating jet lag with a transmeridian flight (12 time zones) westward [27, 28]. A definitive statement regarding the most effective dose of melatonin in jet lag treatment cannot be made; however, separate studies have shown a greater efficacy of a 5 mg immediate-release melatonin dose relative to the group of patients taking 2 mg delayed-release melatonin [24].

Melatonin, as a dietary supplement, is used widely enough but is not an approved treatment for these types of disorders. The reason for this, as a rule, is the lack of sufficient evidence in the form of clinical trials conducted at the appropriate level to evaluate the clinical effects.

2.2 Melatonin in the treatment of insomnia

Insomnia is a pathological condition caused by a variety of endogenous and exogenous factors. Insomnia is characterized primarily by the difficulty of initiating and maintaining sleep, which results in low-quality daily activity. People suffering from chronic insomnia are usually more prone to psychiatric disorders, primarily anxiety-depressive disorders, and cardiovascular diseases [29]. With age, the prevalence of insomnia increases; one of the reasons for this is an involutional decrease in the level of secretion of melatonin [30], a decrease in its concentration with SCN [6]. According to epidemiological studies, 6% of adults in industrialized countries suffer from a chronic form of insomnia [30]. In addition to night manifestations, accompanied by an increase in sleep latency, a decrease in sleep time, low sleep efficiency, and an increase in wakefulness during sleep, daytime manifestations of this disease are also formed, namely, fatigue, decreased short-term memory, decreased mood, headaches, and gastrointestinal disturbances intestinal tract [31].

The architecture of sleep begins to change already in adulthood, while initially a decrease in the duration of slow sleep is observed. The main goals of treating insomnia are to improve the quality of sleep and its duration and also to improve daily activity. As polysomnographic markers used to objectify the effectiveness of therapy insomnia, wake time after sleep onset (WASO), sleep onset latency (SOL), the number of awakenings, and sleep effectiveness. Despite this, polysomnography is an optional research method. Its use is advisable in cases of suspected secondary genesis of insomnia, as well as to exclude other sleep disorders.

According to the questionnaire, patients with insomnia have higher values (more than 7 points) when questioning on the Insomnia Severity Index (ISI) scale. According to the Pittsburgh Sleep Quality Index (PSQI), there may be more than 5 points. The Beck Depression Questionnaire demonstrates at least the presence of minimal signs of a depressive state, reaching values of 10 or more points. To assess the long-term effects of therapy, keeping a sleep diary is one of the objective methods (recommendation level IIB, based on expert consensus).

According to the recommendations of the American Academy of Sleep Medicine (AASM) from 2008, the use of benzodiazepines and a melatonin receptor agonist (ramelteon) is recommended as a therapy for primary insomnia (psychophysiological, idiopathic, and paradoxical forms). At the same time, there are no clear recommendations regarding the order of initiation of therapy with one of the groups of these drugs. The simultaneous use of melatonin and benzodiazepines is acceptable, to reduce the severity of side effects of the latter. It has been shown that agonists of melatonin receptors have a positive effect on the subjective quality of night sleep and their positive therapeutic effect is objectively confirmed by a polysomnographic study. At the same time, the main criteria for the effectiveness of the treatment of insomnia are achieved, namely, a decrease in WASO and SOL by at least 30 minutes, a decrease in the frequency of awakenings, an increase in sleep duration of more than 6 hours, and an increase in sleep efficiency (ratio of sleep time to recording time) to 80% or more [32, 33]. However, given the short half-life of melatonin and melatonin receptor agonists (e.g., ramelteon), the main clinical effects of these drugs are aimed at the treatment of presomic disorders [33]. In this case, immediate-release melatonin has no other effects on the structure of night sleep, except as a decrease in sleep latency. At the same time, there are observations demonstrating, but not explaining, the reason for the increase in the efficiency of activation of MT1 receptors with SCN, which increases their sensitivity to melatonin, which may be the basis of the therapeutic effect in relation to presominal disorders [34].

One of the mechanisms for implementing the hypnotic effect of melatonin can be realized through hormonal stabilization of the limbic system, which is involved in adaptogenic behavior [7, 9].

According to the recommendations of the European Sleep Research Society (ESRS) from 2017, based on a meta-analysis of 109 studies with a total number of patients 13,969 for the period from 2005 to 2016, melatonin and melatonin receptor agonists have shown unequivocal efficacy in the treatment of insomnia (weak recommendation – low-quality evidence). According to the results of individual studies, polysomnographic criteria for the effectiveness of insomnia therapy were achieved, namely, a decrease in sleep latency and an increase in the total sleep time and sleep efficiency [35, 36]. In a number of studies, even a decrease in the number of nocturnal awakenings was noted, which demonstrated effectiveness in relation to intrasomnic disorders. According to these studies, no dependence of the clinical effect on the dose of melatonin used was revealed. A common opinion formed as a result of the analysis of research data is a high safety profile for melatonin.

Melatonin is approved in Europe for the treatment of primary insomnia in adults over the age of 55, with a level of evidence of 1B (level of evidence based on the results of several randomized, placebo-controlled trials) [37].

Studies are demonstrating the effectiveness and perspective use of new forms of melatonin delivery [38]. Modified release tablet formulations with melatonin (MLT) are clinically more useful in initiating and maintaining sleep in elderly insomniacs than those designed for immediate release. The release of MLT from formulation F(nf)2 (nanofiber mats incorporated into 3-layered tablets containing lactose monohydrate both in the upper and lower layers) was found to be in closer alignment with these effects than the other delivery systems [39].

Among healthy children, sleep problems are observed in 20–40% [40] and, among children with impaired development of the nervous system, up to 80% [41, 42]. In pediatric practice, sleep disturbance is most often found among children with autism spectrum disorder (ASD), attention deficit hyperactivity disorder (ADHD), as well as in anxiety or depressive states [43]. Numerous clinical studies have shown the effectiveness of melatonin in the treatment of falling asleep in patients of various age groups, including children with ASD [44] or adolescents suffering from depression [45]. The physiological concentration of melatonin is crucial for the development of cognitive and behavioral functions [46]. A number of studies have demonstrated a causal relationship between a decrease in melatonin levels and the onset of ASD. Forty percent of children with ASD experienced an increase in serotonin while a decrease in melatonin. An increase in the intermediate metabolite of N-acetylserotonin (NAS) was also observed in 47% of patients [47]. One of the reasons for a decrease in the level of melatonin and an increase in the concentration of its precursor may be due to a violation of the activity of hydroxy-indole-O-methyltransferase [46].

Despite the lack of clinical recommendations, the use of delayed-release melatonin is recommended for children with difficulty maintaining sleep, while immediate-release melatonin is recommended for children with difficulty falling asleep [41, 48]. According to individual recommendations (level of evidence C), melatonin should be used as a sleep inducer at a dose of 1–3 mg 30 minutes before bedtime. To obtain chronobiological effects, a melatonin drug should be taken with immediate release 3–4 hours before bedtime at a dose of 0.2–0.5 mg; the maximum dose for children is 3 mg and for adolescents 5 mg [49].

Despite the fact that in a number of studies melatonin has been shown to be effective in treating insomnia in patients with attention ADHD, its effect on cognitive function and behavior in this population of children has not been found [50].

Melatonin has also been shown to be effective in patients with secondary iatrogenic insomnia receiving beta blockers for hypertension [51] as well as in children with attention deficit hyperactivity disorder (level of IA recommendations based on the results of randomized, placebo-controlled clinical trials) [52, 53].

The use of melatonin in pediatric practice is associated with a minimal number of side effects. However, there are reports of undesirable phenomena of mild severity, namely, an increase in the clinical manifestations of nocturnal enuresis, morning drowsiness, and extremely rare insomnia [54].

Thus, according to the main clinical recommendations in the treatment of insomnia, melatonin has a positive effect both on the subjective quality of night sleep and on its objective characteristics. The drug has a high level of evidence of its effectiveness in the long-term therapy of insomnia in patients older than

55 years, associated mainly with the difficulty of falling asleep and the poor quality of night sleep. Ensuring physiological control of the sleep-wake cycle in children with pathology of the development of the nervous system and patients older than 55 years with insomnia is the goal of replacement therapy with melatonin, since in both groups there is a decrease in the secretion of endogenous melatonin during the night [55, 56].

2.3 Melatonin and parasomnia

Parasomnias are undesirable physical or psychological phenomena that usually form at certain stages of sleep, causing a number of clinical manifestations, including the formation of secondary insomnia. Quite often, parasomnia, especially accompanied by motor manifestations, can lead to injuries of varying severity and the formation of psychological problems or social maladaptation [21, 57]. The most striking in its clinical manifestation is REM behavior disorder (RBD). In the treatment of this form of parasomnia, clonazepam is most successfully used. But, the use of this drug is associated with numerous side effects typical of benzodiazepines, especially if the elderly patient has sleep-related breathing disorders (SRBD). An alternative pharmacological method is the use of melatonin. Melatonin also causes a decrease in the frequency and severity of motor activity during an RBD episode, which leads to a decrease in the frequency and severity of injuries. According to the results of a few studies, the use of melatonin at a dose of 3–15 mg led to a significant reduction in paradoxical sleep without atony, as well as the severity of motor manifestations of behavior disorder in the REM phase [58]. One of the options for therapeutic treatment may be taking the drug melatonin for 5–7 days at a minimum dose of 3 mg, followed by an increase in the dose of the drug every 5–7 days to a maximum of 12 mg at night [59, 60]. Little information is available regarding the efficacy of prolonged forms of melatonin or agonists in patients with RBD. There were also no comparisons of the clinical efficacy of clonazepam and melatonin.

Indeed, a number of studies demonstrate a more effective therapeutic effect with the combination of clonazepam and melatonin [61]. The potentiation of the effects of melatonin and clonazepam in the context of RBD therapy has no definitive explanation. It is believed that clonazepam reduces the phase activity inherent in paradoxical sleep, but at the same time, motor activity and minimal disturbance of behavior may remain, according to a polysomnographic survey [62]. The effect of melatonin in combination with clonazepam is due to the modulating effect of the structure of paradoxical sleep, reducing the number of transitions to other stages [59]. An alternative hypothesis explaining the effectiveness of melatonin in RBD may be its effect on increasing the effect of GABA on the GABA receptors of motor neurons of the anterior horns of the spinal cord, which leads to more intense muscle atony. Efficiency may also be related to the fact that melatonin helps to reduce the concentration of calmodulin, which affects the structure of the cytoskeleton and nicotinic acetylcholine receptors of skeletal muscles, which also leads to a progressive decrease in muscle tone [61]. The presence of a favorable safety profile makes the use of melatonin more attractive relative to clonazepam, especially in the elderly [61]. Therefore, in some few clinical trials, melatonin is used as a first-line therapy for RBD, especially in the presence of cognitive impairment, Parkinsonism, or SRBD. In the presence of minimal effectiveness of melatonin or a decrease in its effectiveness during therapy, clonazepam should be additionally prescribed. According to AASM recommendations, melatonin has a "B" level of evidence regarding its effectiveness. Doses of the drug in the studies on the basis of which these recommendations were made ranged from 8 to 12 mg; therefore, there are no clear recommendations regarding the dose of administration [63].

There is also another class of parasomnia in the treatment of which the effectiveness of melatonin was studied. These are parasomnia associated with slow eye movement, which is defined as undesirable motor and psychophysiological manifestations that occur at the time of awakening from a slow-wave sleep. Parasomnia associated with slow eye movement is defined as undesirable motor and psychophysiological manifestations that occur at the time of awakening from a slow-wave sleep [64]. In cases of severe clinical manifestations of these forms of parasomnia, benzodiazepines (clonazepam) or antidepressants (imipramine or clomipramine) may be used. When walking in a dream, the drugs of choice are benzodiazepines or selective serotonin reuptake inhibitors (SSRIs), such as paroxetine and imipramine [64]. The use of melatonin did not reveal a reliable therapeutic effect on the clinical manifestations of these forms of parasomnia. There are only a few studies on the use of melatonin as a first-line therapy for nightly fears in children; the first-line drug is melatonin or L-5-hydroxytryptophan [65]. The absence of a significant clinical effect is associated with the absence of a homeostatic effect on sleep in melatonin.

2.4 Melatonin in the treatment of complications of sleep-dependent respiratory disorders

Sleep-dependent respiratory disorders are represented by several types of pathological conditions: Obstructive sleep apnea (OSA), central sleep apnea, sleep-related hypoventilation, and sleep-related hypoxemia disorder. Most studies are devoted to the study of melatonin metabolism in OSA. A number of studies have demonstrated impaired melatonin secretion in OSA. At the same time, it is believed that the decrease in secretion is secondary. There is also data on the relationship between the concentration of melatonin at night and the duration of night sleep, as well as body weight [66–68]. Some studies have shown a relationship between the severity of OSA and the degree of decrease in melatonin [69]. Approximately 25% of patients with OSA have an altered circadian rhythm of melatonin secretion. In patients with OSA with a maintained rhythm of secretion, peak melatonin levels at night are significantly lower than in healthy people. The 3-month treatment period with continuous positive airway pressure (CPAP) can help restore the physiological rhythm of melatonin in patients with OSA with an impaired secretion profile [70]. One of the uses of melatonin is its use as a drug that reduces the complications associated with respiratory failure during sleep. Numerous studies on biological models demonstrate the positive effect of melatonin on the unfolding pathophysiological cascade of changes in the body in the presence of sleep-dependent respiratory disorders. For example, melatonin inhibits an increase in glucose, the concentration of which increases during periods of apnea [71]. Melatonin modulation of the activity of adenosine monophosphate-activated protein kinase reduces the progression of cardiac muscle hypertrophy. Melatonin also inhibits the expression of inflammatory cytokines, such as tumor necrosis factor alpha, interleukin-6, and cyclooxygenase-2 [72]. It also helps to reduce the severity of Ca^{2+} caused by impaired myocardial contractile function, thus reducing the manifestations of endothelial dysfunction.

The use of melatonin as a prophylactic helps to prevent cardiac remodeling due to hypoxia arising from obstructive apnea [73]. Effects on the cardiovascular system are also realized due to the ability of melatonin and melatonin receptor agonists to inhibit bradykinin B2 receptors, as well as dimerization of angiotensin-converting enzyme I, improving therapeutic control of blood pressure [74]. Another way of realizing the effects of melatonin is the stabilizing effect on angiotensin II receptors and ACE-B2R dimers, which increases the production of nitric oxide by endothelial cells, increasing tissue perfusion. The activation of the MT1 receptor promotes vasoconstriction and MT2 receptor vasodilation. Thus, melatonin can act as a therapeutic agent in the treatment of cardiovascular diseases and hypertension resulting from comorbid diseases in sleep-dependent respiratory disorders. These effects of melatonin in carotid-dependent respiratory disorders were found as a result of a few studies; therefore, they do not have a sufficient recommended level.

2.5 Melatonin in the treatment of hypersomnia

Hypersomnia, such as type I and type II narcolepsy, and idiopathic hypersomnia, are diseases of which the main clinical syndrome is excessive daytime sleepiness. At the same time, drowsiness, being one of the obligate syndromes of diseases, can be modulated by sleep disturbances, observed in these patients, associated with disturbances in sleep structure, and the stability of being in a slow-wave sleep. Currently, drugs approved by FDA, for example, include methylphenidate, modafinil, oxybate, and pitolisant. Methylphenidate, being an analogue of amphetamine, blocks the transport of dopamine and norepinephrine, increasing their concentration. This drug has a fairly large number of side effects. Modaphenyl is better tolerated but may cause psychological dependence on administration [75]. Oxybate and pitolisant are well tolerated. Pitolisant is currently undergoing an expansion of indications up to 6 years of age in the treatment of types 1 and 2 narcolepsy.

Melatonin can affect the severity of hypersomnia in these patients indirectly due to the effect on the architecture of night sleep. A positive impact on the architecture of night sleep is realized by increasing the representation of paradoxical sleep. The positive effects of melatonin administration in patients with hypersomnia in Parkinson's disease have been described, slowing down the decrease in the loss of dopamine-producing neurons and contributing to the suppression of dopamine transport [76]. Presumably, one of the causes of excessive daytime sleepiness in Parkinson's disease is the decrease in the concentration of melatonin [77]. The use of melatonin in patients with neurodegenerative diseases is promising, since a number of interesting effects of melatonin exposure were obtained on biological models. For example, melatonin, freely penetrating the blood-brain barrier, activates brain-derived neurotrophic factor and cyclooxygenase-10, suppressing plasma tumor necrosis factor (TNF-alpha) and IL-10 levels. In experiments, a decrease in the number of apoptotic cells induced by phenylhydrazine was demonstrated. These studies confirm the role of melatonin in neuroprotection and protection against apoptosis in oxidative damage to neurons [78]. According to domestic guidelines for the treatment of nonmotor manifestations of Parkinson's disease, melatonin is recommended for use as a therapy for excessive daytime sleepiness [79].

3. Conclusion

A decrease in the secretion of melatonin is often observed with aging and diseases of various etiologies. Inadequate sleep hygiene, namely, excessive night illumination or night work, are the most common causes of suppression of pineal melatonin production, which has a chronobiological effect on the body. A decrease in the production of melatonin in some cases can be caused by neurodegeneration, accompanied by a change in the functioning of SCN, disrupting the operation of the circadic oscillator. The most common manifestations of epiphyseal deficiency of this hormone are various functional psychopathological disorders in the form of insomnia, anxiety, or depressive disorders. The role of melatonin is currently being actively discussed in the treatment of insomnia and the sleep-wake cycle disorder. A few clinical studies demonstrate the effects in the treatment of the main manifestations of such forms of sleep disorders as hypersomnia and parasomnia. A positive effect is noted in the correction of the pathophysiological cascade arising as a result of hypoxia against the background of sleep-dependent respiratory disorders. Thus, the numerous clinical effects of melatonin demonstrate its universal modulating effect on physiological processes in the body and some common features of the pathogenesis of pathological conditions such as insomnia and circadian rhythm disturbances.

Acknowledgements

We thank Pytin Vasiliy and Poverennova Irina (Samara Medical University).

Conflict of interest

The authors declare no conflict of interest.

Notes/thanks/other declarations

We thank the management and rector Samara Medical University (Kolsanov Alexander) for the opportunity to conduct scientific work.

Author details

Alexander Zakharov* and Elena Khivintseva
Department Neurology and Neurosurgery, Samara State Medical University, Samara, Russia

*Address all correspondence to: zakharov1977@mail.ru

References

[1] Amihaesei IC, Mungiu OC. Main neuroendocrine features and therapy in primary sleep troubles. Revista Medico-Chirurgicală a Societăţii de Medici şi Naturalişti din Iaşi. 2012;**116**(3):862-866

[2] Geoffroy PA, Etain B, Franchi JA, Bellivier F, Ritter P. Melatonin and melatonin agonists as adjunctive treatments in bipolar disorder. Current Pharmacogenomics. 2015;**21**(23):3352-3358. DOI: 10.2174/1381612821666150619093448

[3] Galley HF, Lowes DA, Allen L, Cameron G, Aucott LS, Websteret NR. Melatonin as a potential therapy for sepsis: A phase I dose escalation study and an ex vivo whole blood model under conditions of sepsis. Journal of Pineal Research. 2014;**56**(4):427-438. DOI: 10.1111/jpi.12134

[4] Chang YS, Lin MH, Lee JH, Lee PL, Dai YS, Chu KH, et al. Melatonin supplementation for children with atopic dermatitis and sleep disturbance: A randomized clinical trial. JAMA Pediatrics. 2016;**170**(1):35-42. DOI: 10.1001/jamapediatrics.2015.3092

[5] Ng KY, Leong MK, Liang H, Paxinos G. Melatonin receptors: Distribution in mammalian brain and their respective putative functions. Brain Structure & Function. 2017;**222**(7):2921-2939. DOI: 10.1007/s00429-017-1439-6

[6] Wu YH, Zhou JN, Balesar R, Unmehopa U, Bao A, Jockers R, et al. Distribution of MT1 melatonin receptor immunoreactivity in the human hypothalamus and pituitary gland: Colocalization of MT1 with vasopressin, oxytocin, and corticotropin-releasing hormone. The Journal of Comparative Neurology. 2006;**499**(6):897-910. DOI: 10.1002/cne.21152

[7] Claustal B, Leston J. Melatonin: Physiological effects in humans. Neurochirurgye. 2015;**61**(2-3):77-84. DOI: 10.1016/j.neuchi.2015.03.002

[8] Moller M, Baeres FM. The anatomy and innervation of the mammalian pineal gland. Cell and Tissue Research. 2002;**309**(1):139-150. DOI: 10.1007/s00441-002-0580-5

[9] Cipolla-Neto J, Amaral FG. Melatonin as a hormone: New physiological and clinical insights. Endocrine Reviews. 2018;**39**(6):990-1028. DOI: 10.1210/er.2018-00084

[10] Tan DX, Manchester LC, Sanchez-Barcelo E, Mediavilla MD, Reiter RJ. Significance of high levels of endogenous melatonin in mammalian cerebrospinal fluid and in the central nervous system. Current Neuropharmacology. 2010;**8**(3):162-167. DOI: 10.2174/157015910792246182

[11] Ochoa-Sanchez R, Comai S, Spadoni G, Bedini A, Tarzia G, Gobbi G. Melatonin, selective and non-selective MT1/MT2 receptors agonists: Differential effects on the 24-h vigilance states. Neuroscience Letters. 2014;**561**:156-161. DOI: 10.1016/j.neulet.2013.12.069

[12] Ochoa-Sanchez R, Comai S, Lacoste B, Bambico FR, Dominguez-Lopez S, Spadoni G, et al. Promotion of non-rapid eye movement sleep and activation of reticular thalamic neurons by a novel MT2 melatonin receptor ligand. The Journal of Neuroscience. 2011;**31**(50):18439-18452. DOI: 10.1523/JNEUROSCI.2676-11.2011

[13] Gorfine T, Assaf Y, Goshen-Gottstein Y, Yeshurun Y, Zisapel N. Sleep-anticipating effects of melatonin in the human brain. NeuroImage. 2006;**31**(1):410-418. DOI: 10.1016/j.neuroimage.2005.11.024

[14] Gorfine T, Zisapel N. Late evening brain activation patterns and their

relation to the internal biological time, melatonin, and homeostatic sleep debt. Human Brain Mapping. 2009;**30**(2): 541-552. DOI: 10.1002/hbm.20525

[15] Arbon EL, Knurowska M, Dijk DJ. Randomised clinical trial of the effects of prolonged-release melatonin, temazepam and zolpidem on slow-wave activity during sleep in healthy people. Journal of Psychopharmacology. 2015;**29**(7):764-776. DOI: 10.1177/0269881115581963

[16] Zisapel N. Sleep and sleep disturbances: Biological basis and clinical implications. Cellular and Molecular Life Sciences. 2007;**64**(10):1174-1186. DOI: 10.1007/s00018-007-6529-9

[17] Emens JS, Eastman CI. Diagnosis and treatment of non-24-h sleep-wake disorder in the blind. Drugs. 2017;**77**(6):637-650. DOI: 10.1007/s40265-017-0707-3

[18] Kyba C, Kantermann T. Does ambient light at night reduce total melatonin production? Hormones. 2016;**15**(1):142-143. DOI: 10.14310/horm.2002.1613

[19] Micic G, Lovato N, Gradisar M, Burgess HJ, Ferguson SA, Kennaway DJ, et al. Nocturnal melatonin profiles in patients with delayed sleep-wake phase disorder and control sleepers. Journal of Biological Rhythms. 2015;**30**(5):437-448. DOI: 10.1177/0748730415591753

[20] Leonardo-Mendonca RC, Martinez-Nicolas A, de Teresa G, Ocaña-Wilhelmi J, Rusanova R, Guerra-Hernández E, et al. The benefits of four weeks of melatonin treatment on circadian patterns in resistance-trained athletes. Chronobiology International. 2015;**32**(8):1125-1134. DOI: 10.3109/07420528.2015.1069830

[21] Auger RR, Burgess HJ, Emens JS, Deriy LV, Thomas SM, Sharkey KM. Clinical practice guideline for the treatment of intrinsic circadian rhythm sleep-wake disorders: Advanced sleep-wake phase disorder (ASWPD), delayed sleep-wake phase disorder (DSWPD), non-24-hour sleep-wake rhythm disorder (N24SWD), and irregular sleep-wake rhythm disorder (ISWRD). An update for 2015. Journal of Clinical Sleep Medicine. 2015;**11**(10):1199-1236. DOI: 10.5664/jcsm.5100

[22] Roth T, Nir T, Zisapel N. Prolonged release melatonin for improving sleep in totally blind subjects: A pilot placebo-controlled multicenter trial. Nature and Science of Sleep. 2015;**7**:13-23. DOI: 10.2147/nss.s71838

[23] Suhner A, Schlagenhauf P, Hofer I, Johnson R, Tschopp A, Steffen R. Effectiveness and tolerability of melatonin and zolpidem for the alleviation of jet lag. Aviation, Space, and Environmental Medicine. 2001;**72**:638-646

[24] Suhner A, Schlagenhauf P, Johnson R, Tschopp A, Steffen R. Comparative study to determine the optimal melatonin dosage form for the alleviation of jet lag. Chronobiology International. 1998;**15**(6):655-666. DOI: 10.3109/07420529808993201

[25] Comperatore CA, Lieberman HR, Kirby AW, Adams B, Crowley JS. Melatonin efficacy in aviation missions requiring rapid deployment and night operations. Aviat Space & Environmental Medcine. 1996;**67**(6):520-524

[26] Paul MA, Gray G, Sardana TM, Pigeau RA. Melatonin and zopiclone as facilitators of early circadian sleep in operational air transport crews. Aviation, Space, and Environmental Medicine. 2004;**75**(5):439-443

[27] Petrie K, Conaglen JV, Thompson L, Chamberlain K. Effect of melatonin on jet lag after long haul flights. BMJ.

1989;**298**(6675):705-707. DOI: 10.1136/bmj.298.6675.705

[28] Petrie K, Dawson AG, Thompson L, Brook R. A double-blind trial of melatonin as a treatment for jet lag in international cabin crew. Biological Psychiatry. 1993;**33**(7):526-530. DOI: 10.1016/0006-3223(93)90007-z

[29] Hoevenaar-Blom MP, Spijkerman AM, Kromhout D, van den Berg JF, Verschuren WM. Sleep duration and sleep quality in relation to 12-year cardiovascular disease incidence: The MORGEN study. Sleep. 2011;**34**(11):1487-1492. DOI: 10.5665/sleep.1382

[30] Zhang W, Wing YK. Sex differences in insomnia: A meta-analysis. Sleep. 2006;**29**(1):85-93. DOI: 10.1093/sleep/29.1.85

[31] Edinger JD, Bonnet MH, Bootzin RR, Doghramji K, Dorsey CM, Espie CA, et al. Derivation of research diagnostic criteria for insomnia: Report of an American Academy of Sleep Medicine Work Group. Sleep. 2004;**27**(8):1567-1596. DOI: 10.1093/sleep/27.8.1567

[32] Schutte-Rodin S, Broch L, Buysse D, Dorsey C, Sateia M. Clinical guideline for the evaluation and management of chronic insomnia in adults. Journal of Clinical Sleep Medicine. 2008;**4**(5):487-504. DOI: 10.5664/jcsm.27286

[33] Erman M, Seiden D, Zammit G, Sainati S, Zhang J. An efficacy, safety, and dose-response study of Ramelteon in patients with chronic primary insomnia. Sleep Medicine. 2006;**7**(1):17-24. DOI: 10.1016/j.sleep.2005.09.004

[34] Dubocovich ML. Melatonin receptors: Role on sleep and circadian rhythm regulation. Sleep Medicine. 2007;**8**(3):34-42. DOI: 10.1016/j.sleep.2007.10.007

[35] Brzezinski A, Vangel MG, Wurtman RJ, Norrie G, Zhdanova I, Ben-Shushan A, et al. Effects of exogenous melatonin on sleep: A meta-analysis. Sleep Medicine Reviews. 2005;**9**(1):41-50. DOI: 10.1016/j.smrv.2004.06.004

[36] Zakharov AV, Khivintseva EV, Pyatin VF, Sergeeva MS, Antipov OI. Melatonin—Known and novel areas of clinical application. Neuroscience and Behavioral Physiology. 2019;**49**(1):60-63. DOI: 10.1007/s11055-018-0692-3

[37] Wilson SJ, Nutt DJ, Alford C, Argyropoulos SV, Baldwin DS, Baldwin DS, et al. British Association for Psychopharmacology consensus statement on evidence-based treatment of insomnia, parasomnias and circadian rhythm disorders. Journal of Psychopharmacology. 2010;**24**(11):1577-1601. DOI: 10.1177/0269881110379307

[38] Vlachou M, Kikionis S, Siamidi A, Tragou K, Ioannou E, Roussis V, et al. Modified in vitro release of melatonin loaded in nanofibrous electrospun mats incorporated into mono-layered and three-layered tablets. Journal of Pharmaceutical Sciences. 2019;**108**(2):970-976. DOI: 10.1016/j.xphs.2018.09.0351

[39] Vlachou M, Tragou K, Siamidi A, Kikionis S, Chatzianagnostou AL, Mitsopoulos A, et al. Modified in vitro release of the chronobiotic hormone melatonin from matrix tablets based on the marine sulfated polysaccharide ulvan. Journal of Drug Delivery Science and Technology. 2018;**44**:41-48. DOI: 10.1016/j.jddst.2017.11.019

[40] Fricke-Oerkermann L, Plück J, Schredl M, Heinz K, Mitschke A, Wiater A, et al. Prevalence and course of sleep problems in childhood. Sleep. 2007;**30**(10):1371-1377. DOI: 10.1093/sleep/30.10.1371

[41] Grigg-Damberger M, Ralls F. Treatment strategies for complex

behavioral insomnia in children with neurodevelopmental disorders. Current Opinion in Pulmonary Medicine. 2013;**19**(6):616-625. DOI: 10.1097/mcp.0b013e328365ab89

[42] Damiani JM, Sweet BV, Sohoni P. Melatonin: An option for managing sleep disorders in children with autism spectrum disorder. American Journal of Health-System Pharmacy. 2014;**71**(2):95-101. DOI: 10.2146/ajhp130215

[43] Meltzer LJ, Mindell JA. Sleep and sleep disorders in children and adolescents. Psychiatric Clinics of North America. 2006;**29**(4):1059-1076. DOI: 10.1016/j.psc.2006.08.004

[44] Goldman SE, Adkins KW, Calcutt MW, Carter MD, Goodpaster RL, Wang L, et al. Melatonin in children with autism spectrum disorders: Endogenous and pharmacokinetic profiles in relation to sleep. Journal of Autism and Developmental Disorders. 2014;**44**(10):2525-2535. DOI: 10.1007/s10803-014-2123-9

[45] Bartlett DJ, Biggs SN, Armstrong SM. Circadian rhythm disorders among adolescents: Assessment and treatment options. The Medical Journal of Australia. 2013;**199**:16-20. DOI: 10.5694/mja13.10912

[46] Melke J, Goubran Botros H, Chaste P, Betancur C, Nygren G, Anckarsäter H, et al. Abnormal melatonin synthesis in autism spectrum disorders. Molecular Psychiatry. 2008;**13**:90-98. DOI: 10.1038/sj.mp.4002016

[47] Pagan C, Delorme R, Callebert J, Goubran-Botros H, Amsellem F, Drouot X, et al. The serotonin-N-acetylserotonin–melatonin pathway as a biomarker for autism spectrum disorders. Translational Psychiatry. 2014;**4**:479. DOI: 10.1038/tp.2014.120

[48] Veatch OJ, Goldman SE, Adkins KW, Malow BA. Melatonin in children with autism spectrum disorders: How does the evidence fit together? Journal of Nature and Science. 2015;**1**(7):125

[49] Bruni O, Alonso-Alconada D, Besag F, Biran V, Braam W, Cortese S, et al. Current role of melatonin in pediatric neurology: Clinical recommendations. European Journal of Paediatric Neurology. 2015;**19**:122-133. DOI: 10.1016/j.ejpn.2014.12.007

[50] Van der Heijden KB, Smits MG, Van Someren EJ, Ridderinkhof KR, Gunning WB. Effect of melatonin on sleep, behavior, and cognition in ADHD and chronic sleep-onset insomnia. Journal of the American Academy of Child and Adolescent Psychiatry. 2007;**46**:233-241. DOI: 10.1097/01.chi.0000246055.76167.0d

[51] Scheer FA, Morris CJ, Garcia JI, Smales C, Kelly EE, Marks J, et al. Repeated melatonin supplementation improves sleep in hypertensive patients treated with beta-blockers: A randomized controlled trial. Sleep. 2012;**35**:1395-1402. DOI: 10.5665/sleep.2122

[52] Shechter A, Lesperance P, Ng YKN, Boivin DB. Nocturnal polysomnographic sleep across the menstrual cycle in premenstrual dysphoric disorder. Sleep Medicine. 2012;**13**:1071-1078. DOI: 10.1016/j.sleep.2012.05.012

[53] Holvoet E, Gabriels L. Disturbed sleep in children with ADHD: Is there a place for melatonin as a treatment option? Tijdschrift voor Psychiatrie. 2013;**55**:349-357

[54] Andersen IM, Kaczmarska J, McGrew SG, Malow BA. Melaton in for insomniain children with autism spectrum disorders. Journal of Child Neurology. 2008;**23**:482-485. DOI: 10.1177/0883073807309783

[55] Haimov I, Laudon M, Zisapel N, Souroujon M, Nof D, Shlitner A, et al. Sleep disorders and melatonin rhythms in elderly people. BMJ. 1994;**309**(6948):167. DOI: 10.1136/bmj.309.6948.167

[56] Tordjman S, Najjar I, Bellissant E, Anderson GM, Barburoth M, Cohen D, et al. Advances in the research of melatonin in autism spectrum disorders: Literature review and new perspectives. International Journal of Molecular Sciences. 2013;**14**:20508-20542. DOI: 10.3390/ijms141020508

[57] Zakharov AV, Poverennova IE, Kalinin VA, Khivintseva EV. Parasomnias associated with disordered arousal from slow-wave sleep: Mechanism of occurrence and neurophysiological characteristics. Neuroscience and Behavioral Physiology. 2020;**50**:270-274. DOI: 10.1007/s11055-020-00897-z

[58] Kunz D, Mahlberg R. A two-part, double-blind, placebocontrolled trial of exogenous melatonin in REM sleep behaviour disorder. Journal of Sleep Research. 2010;**19**:591-596. DOI: 10.1111/j.1365-2869.2010.00848.x

[59] Kunz D, Bes F. Melatonin as a therapy in rem sleep behavior disorder patients: An open-labeled pilot study on the possible influence of melatonin on rem-sleep regulation. Movement Disorders. 1999;**14**:507-511. DOI: 10.1002/1531-8257(199905)14:3<507::aid-mds1021>3.0.co;2-8

[60] Schaefer C, Kunz D, Bes F. Melatonin effects in REM sleep behavior disorder associated with obstructive sleep apnea syndrome: A case series. Current Alzheimer Research. 2017;**14**:1084-1089. DOI: 10.2174/1567205014666170523094938

[61] McGrane IR, Leung JG, St Louis EK, Boeve BF. Melatonin therapy for REM sleep behavior disorder: A critical review of evidence. Sleep Medicine. 2015;**16**:19-26. DOI: 10.1016/j.sleep.2014.09.011

[62] Lapierre O, Montplaisir J. Polysomnographic features of REM sleep behavior disorder development of a scoring method. Neurology. 1992;**42**:1371. DOI: 10.1212/wnl.42.7.1371

[63] Aurora RN, Zak RS, Maganti RK, Auerbach SH, Casey KR, Chowdhuri S, et al. Best practice guide for the treatment of REM sleep behavior disorder (RBD). Journal of Clinical Sleep Medicine. 2010;**6**(1):85-95

[64] Sateia MJ. International classification of sleep disorders-third edition. Chest. 2014;**146**(5):1387-1394. DOI: 10.1378/chest.14-0970

[65] Bruni O, Ferri R, Miano S, Verrillo E. L-5-Hydroxytryptophan treatment of sleep terrors in children. European Journal of Pediatrics. 2004;**163**:402-407. DOI: 10.1007/s00431-004-1444-7

[66] Wetterberg L, Bratlid T, Knorring L, et al. A multinational study of the relationships between nighttime urinary melatonin production, age, gender, body size and latitude. European Archives of Psychiatry and Clinical Neuroscience. 1999;**249**:256-262. DOI: 10.1007/s004060050095

[67] Scheer F, Czeisler C. Melatonin, sleep and circadian rhythms. Sleep Medicine Reviews. 2005;**9**(1):5-9. DOI: 10.1016/j.smrv.2004.11.004

[68] Zhdanova I, Tucci V. Melatonin, circadian rhythms and sleep. Current Treatment Options in Neurology. 2003;**5**:225-229. DOI: 10.1007/s11940-003-0013-0

[69] Reutrakul S, Siwasaranond N, Nimitphong H, Saetung S, Chirakalwasan N, Chailurkit LO, et al.

Associations between nocturnal urinary 6-sulfatoxymelatonin, obstructive sleep apnea severity and glycemic control in type 2 diabetes. Chronobiology International. 2017;**34**:382-392. DOI: 10.1080/07420528.2016.1278382

[70] Barnaś M, Maskey-Warzęchowska M, Bielicki P, Kumor M, Chazan R. Diurnal and nocturnal serum melatonin concentrations after treatment with continuous positive airway pressure in patients with obstructive sleep apnea. Polish Archives of Internal Medicine. 2017;**127**(9):589-596. DOI: 10.20452/pamw.4062

[71] Kaminski RS, Martinez D, Fagundes M, Martins EF, Montanari CC, Rosa DP, et al. Melatonin prevents hyperglycemia in a model of sleep apnea. Archives of Endocrinology and Metabolism. 2015;**59**:66-70. DOI: 10.1590/2359-3997000000012

[72] Xie S, Deng Y, Pan YY, Wang ZH, Ren J, Guo XL, et al. Melatonin protects against chronic intermittent hypoxia-induced cardiac hypertrophy by modulating autophagy through the 5′ adenosine monophosphate-activated protein kinase pathway. Biochemical and Biophysical Research Communications. 2015;**464**:975-981. DOI: 10.1016/j.bbrc.2015.06.149

[73] Yeung HM, Hung MW, Lau CF, Fung ML. Cardioprotective effects of melatonin against myocardial injuries induced by chronic intermittent hypoxia in rats. Journal of Pineal Research. 2015;**58**:12-25. DOI: 10.1111/jpi.12190

[74] Sabatini RA, Guimaraes PB, Fernandes LB, Reis FC, Bersanetti PA, Mori MA, et al. ACE activity is modulated by kinin B2 receptor. Hypertension. 2008;**51**(3):689-695. DOI: 10.1161/HYPERTENSIONAHA.107.091181

[75] Roth T, Schwartz JR, Hirshkowitz M, Erman MK, Dayno JM, Arora S. Evaluation of the safety of modafinil for treatment of excessive sleepiness. Journal of Clinical Sleep Medicine. 2007;**3**:595-602

[76] Lin CH, Huang JY, Ching CH, Chuang JI. Melatonin reduces the neuronal loss, downregulation of dopamine transporter, and upregulation of D2 receptor in rotenone-induced parkinsonian rats. Journal of Pineal Research. 2008;**44**(2):205-213. DOI: 10.1111/j.1600-079X.2007.00510.x

[77] Videnovic A, Noble C, Reid KJ, Peng J, Turek FW, Marconi A, et al. Circadian melatonin rhythm and excessive daytime sleepiness in Parkinson disease. JAMA Neurology. 2014;**71**:463-469. DOI: 10.1001/jamaneurol.2013.6239

[78] Pazar A, Kolgazi M, Memisoglu A, Bahadir E, Sirvanci S, Yaman A, et al. The neuroprotective and anti-apoptotic effects of melatonin on hemolytic hyperbilirubinemia-induced oxidative brain damage. Journal of Pineal Research. 2016;**60**(1):74-83. DOI: 10.1111/jpi.12292

[79] Srinivasan V, Cardinali DP, Srinivasan US, Kaur C, Brown GM, Spence D, et al. Therapeutic potential of melatonin and its analogs in Parkinson's disease: Focus on sleep and neuroprotection. Therapeutic Advances in Neurological Disorders. 2011;**4**(5):297-317. DOI: 10.1177/1756285611406166

Chapter 3

Melatonin as a Food Supplement for Sleep Disorders

Ioulia K. Tseti

Abstract

N-acetyl-5-methoxytryptamine commonly known as melatonin is a neurohormone produced in the pineal gland as a response to decrease in ambient light and regulates the sleep-wake cycle. Melatonin is a derivative of the amino acid tryptophan and is produced in humans and other mammals. Melatonin supplements are used to treat insomnia and sleep disorders and to adjust sleep schedules due to jet lag. Synthetic melatonin is available as a food supplement in various dosage forms such as pills, granules for oral solution, orodispersible granules, and syrups in order to address patients' needs. Melatonin is often combined with water-soluble vitamins such as B complex vitamins and minerals like zinc in order to be more effective.

Keywords: melatonin, food supplements, jet lag, sleeping disorder, dosage forms

1. Introduction

In 1960, the chemical structure of melatonin was defined [1] since significant attention was attracted towards its use a few years earlier when the dermatologist Dr. Lerner and his colleagues observed that melatonin could cause the lightening of frog skin [2]. Melatonin has been found to affect a wide range of physiological processes such as sleep-wake cycles [3], circadian rhythms [4], sexual maturation [5], and aging [6].

Since then, exogenous melatonin has demonstrated a series of clinical effects [7, 8], and numerous clinical studies have been conducted, where improved sleep quality was documented following exogenous melatonin administration [9]. Recent studies demonstrated analgesic [10], anxiolytic [11], anti-inflammatory, and anti-oxidative effects [12] following administration of melatonin.

2. Melatonin as a food supplement and its uses in modern life

Melatonin is synthesized from tryptophan via 5-hydroxytryptophan and 5-hydroxytryptamine (serotonin). This is followed by *N*-acetylation of serotonin by *N*-acetyltransferase (arylalkylamine *N*-acetyltransferase, AA-NAT) to *N*-acetylserotonin (NAT) and *O*-methylation by acetylserotonin *O*-methyltransferase (ASMT) [previously known as hydroxyindole-*O*-methyltransferase (HIOMT)] to melatonin (*N*-acetyl-5-methoxytryptamine). The rhythm of melatonin production is endogenous, being generated by clock genes in the suprachiasmatic nuclei (SCN), the major central rhythm-generating system or "clock" in mammals. The rhythm, as for the circadian system in general, is synchronized to 24 h primarily by the light-dark cycle acting via the retina and the retinohypothalamic projection to the SCN [13].

In humans melatonin is metabolized, 70% to 6-sulphatoxy melatonin (aMT6s), primarily within the liver, by 6-hydroxylation, followed by sulfate conjugation; this mechanism varies through species. A number of minor metabolites are also formed, including the glucuronide conjugate. *N1*-acetyl-*N2*-formyl-5-methoxykynuramine and *N1*-acetyl-5-methoxykynuramine were initially reported as brain metabolites [14, 15], but were proved difficult to detect in plasma or urine except after administration of exogenous melatonin [16]. Exogenous oral fast release or intravenous melatonin has a short metabolic half-life, i.e., 20–60 min, depending on the species—with a large hepatic first-pass effect and a biphasic elimination pattern [17]. Slow release, prolonged release, and surge sustained preparations are designed to extend the time of high circulating melatonin [18]. Melatonin has low bioavailability, in general, although it has been found that transmucosal administration increases bioavailability [19]. A critical feature of exogenous melatonin with regard to its clinical uses is its very low toxicity and lack of addictive properties [20, 21].

A T_{max} of approximately 50 min has been reported following oral immediate-release formulation of melatonin. T1/2 of both oral and intravenous melatonin was about 45 min [22]. Over 80% of melatonin dose is excreted exclusively in the urine, as 6-sulfatoxymelatonin (6-SMT) following first-pass hepatic metabolism [23, 24]. Melatonin is short-lived in humans with a half-life in plasma of only 40–50 min [23]. Following oral administration, it is rapidly absorbed with peak plasma levels occurring between 20 min and 2 h depending on dose [14].

Administration of melatonin 45 min before intended clinical effect may therefore be recommended. However, external factors, such as caffeine [25], smoking [26], and other medications [27, 28], which may potentially affect the pharmacokinetics of melatonin, should be considered prior to exogenous melatonin administration.

Long-term use of sedative-hypnotics for insomnia lacks evidence of treatment and has traditionally been discouraged for reasons that include concerns about potential adverse drug effects, such as cognitive impairment, daytime sedation, motor incoordination, and risk of motor vehicle accidents and slips and falls. In addition, the effectiveness and safety of long-term use of these agents remain to be determined [29]. Moreover, several studies have been conducted to assess the effects of sleep hygiene interventions and various non-pharmacological interventions, such as physical activity, bright light exposure, and noise abatement, but no definite effect on nighttime sleep has been reported [29]. Many people seek treatment for insomnia using alternative and complementary medicine [30]. Generally, the main goal of non-pharmacological remedies in the treatment of primary insomnia is to correct behavior patterns that are not conducive to a good quality sleep, and nutrients might play a significant role in this setting, but no evidence is available as to the preferred alternative treatment of insomnia.

In addition to melatonin, other micronutrients, such as zinc and magnesium, may play a role in facilitating sleep. Zinc exhibits an antidepressant-like activity, as observed in a preclinical model of depression [31–34]. Significant clinical correlates were shown [35] related to its action as an antagonist of the glutamate/*N*-methyl-*D*-aspartate receptor. Magnesium has beneficial effects on mood and is crucial, together with zinc, in the endogenous synthesis of melatonin [36]. Various food supplements contain combinations of melatonin with either magnesium or zinc.

Since melatonin production declines with age and is lower in middle-aged and elderly adults with insomnia than in good sleepers [37], supplementation with exogenous melatonin is very common. Exogenous melatonin can effectively treat insomnia by mimicking the natural endogenous melatonin, binding to the same receptors, and activating the same downstream pathways. The effect of melatonin on sleep is believed to be a consequence of mechanisms that involve an increase in sleep propensity by enhancing the amplitude of circadian clock oscillations via melatonin type 1 (MT1) receptors and the synchronization of the circadian clock via melatonin type 1 (MT2) receptors [39]. By activating MT1 (melatonin type 1 receptor) and MT2 (melatonin type 2 receptor) receptors, melatonin and nonselective MT1/MT2 receptor agonists have shown to improve sleep quality, increase total sleep time, improve sleep efficiency, and decrease sleep onset latency in insomnia patients [38].

The mammalian circadian clock covers a wide range of physiological processes and plays pivotal role in reproduction [40, 41]. It is currently accepted that dysregulation of the circadian rhythm caused by night shifts, jet lag, and sleep deprivation has a detrimental effect on the reproductive system [42, 43]. Melatonin is produced not only by the pineal gland but also in glial cells, meningeal cells, and in other peripheral tissues, and its cyclical pattern of secretion is responsive to zeitgebers [44]. Melatonin permeability into the central nervous system was described decades ago [45], and its efficient transport through the blood-brain barrier promotes accumulation in the brain at levels higher than the ones existing in the blood. Melatonin also possesses neuroprotective and antioxidant properties [42]. Modulation of redox signaling systems influences the reproductive system in both animals and humans [43, 46], and it is known that insufficient endogenous production of melatonin has been associated with disturbances in the reproductive system due to increased levels of reactive oxygen species (ROS), which are harmful to the male and female gametes [47]. Unhealthy lifestyles and psychosocial stress are aspects of modern life that have a negative impact on gynecological health and reproduction [48]. Epidemiological studies show that night shifts may negatively influence fetal development and may exacerbate gynecological and metabolic disorders, including endometriosis, diabetes, and obesity [49]. Consequently, melatonin supplementation has been considered as a therapeutic approach in gynecological practice owing to its antioxidant properties and its action as hormone modulator.

The neurohormone melatonin is not stored in the pineal gland, but rather is released into the bloodstream and can penetrate all body tissues [50]. It is important to note that "darkness" stimulates the pineal gland to secrete melatonin, whereas exposure to light inhibits this mechanism [51].

Regarding the actual administration of melatonin, it has been shown that the timing of melatonin administration is more crucial in producing the best results than the actual dose; this is secondary to the normal physiologic function of the circadian rhythm [51]. It has been reported that when melatonin was administered at bedtime as a "sleeping pill," it was not effective unless high doses were used [52]; however, when small doses of melatonin were administered to patients about 2–4 h before bedtime, it was shown to be effective in decreasing sleep latency [53].

Garfinkel et al. [54] investigated 12 elderly subjects (mean age 76 ± 8 years) with chronic illness and insomnia in a crossover study using wrist actigraphy comparing administration of PR-melatonin for 3 weeks with placebo. PR-melatonin 2 mg produced a statistically significant improvement in sleep efficiency and wake time after sleep onset was shorter. Sleep latency decreased, but this was not statistically significant, while total sleep time was not affected.

The side effect profile of melatonin therapy is quite reassuring and is largely superior to other sleep-inducing agents. For example, melatonin therapy does not cause withdrawal or dependence symptoms, unlike benzodiazepines (BZDs) and z-drugs such as zolpidem [55].

Potential harmful effects of exogenous melatonin therapy might result in amenorrhea when used in large doses, which is likely due to suppression of gonadotropin-releasing hormone (GnRH) [56]. However, this effect is readily reversible with cessation of the medication.

Sleep disorders, regardless of the etiology, are frequently encountered by physicians and other health-care providers. According to data from the Centers for Disease Control and Prevention (CDC), up to about 70 million Americans suffer from chronic sleep problems [57], while according to the American Psychiatric Association (APA), approximately 30% of all adults suffer from sleep disorders [58]. Considering that, supplements containing melatonin are widely used. Over the counter (OTC) melatonin-containing supplements can be easily found either online or at pharmacy stores, with beneficial claims on jet lag [59], as well as occasional sleepiness, sleep problems caused by stress, overall mood, and overall health.

Melatonin is often combined with vitamins, such as B complex vitamins and micronutrients, i.e., zinc or magnesium. Clinical studies have been conducted demonstrating the synergistic effect of these combinations [60]. Magnesium supplementation improves sleep efficiency, sleep time and sleep onset latency, early morning awakening, and insomnia objective measures such as the concentration of serum renin, melatonin, and serum cortisol, in older adults [61]. Meanwhile, there is clear evidence on the antidepressant effect of vitamin B12 [62] and vitamin B6 for therapy of hormone-related depression in women [8].

OTC melatonin food supplements are supplied in various pharmaceutical dosage forms in order to accommodate all patients' needs. Usually these food supplements contain 1 mg of melatonin in order to be able to bear the EFSA claim of "Melatonin contributing to the reduction of time taken to fall asleep and to the alleviation of subjective feelings of jet lag" [34].

Pharmaceutical dosage forms of melatonin-containing supplements include tablets and granules either for direct administration or for oral solution preparations, while recently the push and drink form is becoming popular. In this dosage form, the solid mixture containing melatonin and other ingredients is airtightly separated from the solution used for dissolution of the solid. This is achieved by using a container based on a closing storage cap-vial system, in which the closing storage cap contains the solid composition, while the vial contains the solution composition. The nutritional supplement is prepared just prior to use by an immediate procedure, and it can be consumed directly from the vial.

3. Conclusions

Since melatonin has a very low side effect profile and limited evidence of habituation and tolerance, it is widely used among people that suffer from sleep disorders. Various clinical trials have been conducted proving the efficacy of melatonin in treating sleep disorders regardless of the etiology. A plethora of OTC melatonin-containing food supplements, displayed in various pharmaceutical dosage forms, is nowadays available covering the needs of the patients.

Conflict of interest

The authors declare no conflict of interest.

Author details

Ioulia K. Tseti
Uni-Pharma Pharmaceutical Laboratories SA, Athens, Greece

*Address all correspondence to: meristoudi@uni-pharma.gr

References

[1] Lerner AB, Case JD, Takahashi Y. Isolation of melatonin and 5-methoxyindole-3-acetic acid from bovine pineal glands. The Journal of Biological Chemistry. 1960;**235**:1992-1997

[2] Lerner AB, Case JD, Takahashi Y, Lee TH, Mori W. Isolation of melatonin, the pineal gland factor that lightens melanocytes. Journal of the American Chemical Society. 1958;**80**(10):2587

[3] Reiter RJ. Pineal melatonin: Cell biology of its synthesis and of its physiological interactions. Endocrine Reviews. 1991;**12**(2):151-180

[4] Gillette MU, Tischkau SA. Suprachiasmatic nucleus: The brain's circadian clock. Recent Progress in Hormone Research. 1999;**54**:33-58. discussion 58-59

[5] Díaz López B, Díaz Rodríguez E, Urquijo C, Fernández ÁC. Melatonin influences on the neuroendocrine-reproductive axis. Annals of the New York Academy of Sciences. 2005;**1057**:337-364

[6] Pierpaoli W, Regelson W. Pineal control of aging: Effect of melatonin and pineal grafting on aging mice. Proceedings of the National Academy of Sciences of the United States of America. 1994;**91**(2):787-791

[7] Brzezinski A. Melatonin in humans. N Engl J Med. 1997;**336**:186-195

[8] Andersen LPH, Werner MU, Rosenberg J, Gögenur I. A systematic review of peri-operative melatonin. Anaesthesia. 2014;**69**:1163-1171

[9] Bourne RS, Mills GH, Minelli C. Melatonin therapy to improve nocturnal sleep in critically ill patients: Encouraging results from a small randomized controlled trail. Critical Care. 2008;**12**:R52

[10] Borazan H, Tuncer S, Yalcin N, Erol A, Otelcioglu S. Effects of preoperative oral melatonin medication on postoperative analgesia, sleep quality, and sedation in patients undergoing elective prostatectomy: A randomized clinical trial. Journal of Anesthesia. 2010;**24**:155-160

[11] Caumo W, Torres F, Moreia NL, Auzani JAS, Monteiro CA, Londero G, et al. The clinical impact of preoperative melatonin on postoperative outcomes in patients undergoing abdominal hysterectomy. Anesthesia and Analgesia. 2007;**105**:1263-1271

[12] Kücükakin B, Lykkesfeldt J, Nielsen HJ, Reiter RJ, Rosenberg J, Gögenur I. Utility of melatonin to treat surgical stress after major vascular surgery—A safety study. Journal of Pineal Research. 2008;**44**:426-431

[13] Klein DC. Photoneural regulation of the mammalian pineal gland. Ciba Foundation Symposium. 1985;**117**: 38-56

[14] Arendt J. Melatonin and the Mammalian Pineal Gland. London: Chapman Hall; 1995

[15] Hirata F, Hayaishi O, Tokuyama T, Seno S. In vitro and in vivo formation of two new metabolites of melatonin. The Journal of Biological Chemistry. 1974;**249**:1311-1313

[16] Ma X, Idle JR, Krausz KW, Tan DX, Ceraulo L, Gonzalez FJ. Urinary metabolites and antioxidant products of exogenous melatonin in the mouse. Journal of Pineal Research. 2006;**40**:343-349

[17] Lane EA, Moss HB. Pharmacokinetics of melatonin in man: First pass hepatic metabolism. The Journal of Clinical Endocrinology and Metabolism. 1985;**61**:1214-1216

[18] Rajaratnam SM, Dijk DJ, Middleton B, Stone BM, Arendt J. Melatonin phase-shifts human circadian rhythms with no evidence of changes in the duration of endogenous melatonin secretion or the 24-hour production of reproductive hormones. The Journal of Clinical Endocrinology and Metabolism. 2003;**88**:4303-4309

[19] Zetner D, Andersen LP, Rosenberg J. Pharmacokinetics of alternative administration routes of melatonin: A systematic review. Drug Research. 2016;**66**:169-173

[20] Sugden D. Psychopharmacological effects of melatonin in mouse and rat. The Journal of Pharmacology and Experimental Therapeutics. 1983;**227**:587-591

[21] Guardiola-Lemaître B. Toxicology of melatonin. Journal of Biological Rhythms. 1997;**12**:697-706

[22] Harpsøe NG, Andersen LPH, Gögenur I, Rosenberg J. Clinical pharmacokinetics of melatonin: A systematic review. European Journal of Clinical Pharmacology. 2015;**71**:901

[23] Waldhauser F, Waldhauser M, Lieberman HR, et al. Bioavailability of oral melatonin in humans. Neuroendocrinology. 1984;**39**(4):307-1325

[24] Arendt J, Skene DJ. Melatonin as a chronobiotic. Sleep Medicine Reviews. 2005;**9**(1):25-39

[25] Härtter S, Nordmark A, Rose DM, Bertilsson L, Tybring G, Laine K. Effects of caffeine intake on the pharmacokinetics of melatonin, a probe drug for CYP1A2 activity. British Journal of Clinical Pharmacology. 2003;**56**:679-682

[26] Ursing C, Bhar CV, Brismar K, Röjdmark S. Influence of cigarette smoking on melatonin levels in man. European Journal of Clinical Pharmacology. 2005;**61**:197-201

[27] Hilli J, Korhonen T, Turpeinen M, Hokkanen J, Mattila S, Laine K. The effect of oral contraceptives on the pharmacokinetics of melatonin in healthy subjects with CYP1A2 g.-163C>A polymorphism. Journal of Clinical Pharmacology. 2008;**48**:986-994

[28] Härrter S, Grözinger M, Wiegmann H, Röschke J, Hiemke C. Increased bioavailability of oral melatonin after fluvoxamine coadministration. Clinical Pharmacology and Therapeutics. 2000;**67**:1-6

[29] Bain KT. Management of chronic insomnia in elderly persons. The American Journal of Geriatric Pharmacotherapy. 2006;**4**:168-192

[30] Chung KF, Lee CK. Over-the-counter sleeping pills: A survey of use in Hong Kong and a review of their constituents. General Hospital Psychiatry. 2002;**24**:430-435

[31] Kroczka B, Zieba A, Dudek D, et al. Zinc exhibits an antidepressant-like effect in the forced swimming test in mice. Polish Journal of Pharmacology. 2000;**52**:403-406

[32] Nowak G, Siwek M, Dudek D, et al. Effect of zinc supplementation on antidepressant therapy in unipolar depression: A preliminary placebo-controlled study. Polish Journal of Pharmacology. 2003;**55**:1143-1147

[33] Siwek M, Dudek D, Paul IA, et al. Zinc supplementation augments efficacy of imipramine in treatment resistant patients: A double blind, placebo-controlled study. Journal of Affective Disorders. 2009;**118**:187-195

[34] Laires MJ, Monteiro CP, Bicho M. Role of cellular magnesium in health

and human disease. Frontiers in Bioscience. 2004;**9**:262-276

[35] Sowa-Kuc'ma M, Legutko B, Szewczyk B, et al. Antidepressant-like activity of zinc: Further behavioral and molecular evidence. Journal of Neural Transmission. 2008;**115**:1621-1628

[36] Touitou Y, Bogdan A. Promoting adjustment of the sleep-wake cycle by chronobiotics. Physiology & Behavior. 2007;**90**:294-300

[37] Haimov I, Laudon M, Zisapel N, et al. Sleep disorders and melatonin rhythms in elderly people. BMJ. 1994;**309**:167

[38] Reppert SM, Weaver DR. Molecular analysis of mammalian circadian rhythms. Annual Review of Physiology. 2001;**63**:647-676

[39] Gamble KL, Resuehr D, Johnson CH. Shift work and circadian dysregulation of reproduction. Frontiers in Endocrinology. 2013;**4**:1-10

[40] Goldman BD. The circadian timing system and reproduction in mammals. Steroids. 1999;**64**:679-685

[41] Beaver L, Gvakharia B, Vollintine T, Hege D, Stanewsky R, Giebultowicz JM. Loss of circadian clock function decreases reproductive fitness in males of *Drosophila melanogaster*. Proceedings of the National Academy of Sciences of the United States of America. 2002;**99**:2134-2139

[42] Genario R, Giacomini AC, Demin KA, dos Santos BE, Marchiori NI, Volgin AD, et al. The evolutionarily conserved role of melatonin in CNS disorders and behavioral regulation: Translational lessons from zebrafish. Neuroscience and Biobehavioral Reviews. 2019;**99**:117-127

[43] Rocha CS, Rato L, Martins AD, Alves MG, Oliveira PF. Melatonin and male reproductive health: Relevance of darkness and antioxidant properties. Current Molecular Medicine. 2015;**15**:299-311

[44] Tan DX, Manchester LC, Fuentes-Broto L, Paredes SD, Reiter RJ. Significance and application of melatonin in the regulation of brown adipose tissue metabolism: Relation to human obesity. Obesity Reviews. 2011;**12**:167-188

[45] Pardridge WM, Mietus LJ. Transport of albumin-bound melatonin through the blood-brain barrier. Journal of Neurochemistry. 1980;**34**:1761-1763

[46] Herxheimer A, Petrie KJ. Melatonin for the prevention and treatment of jet lag. Cochrane Database Systematic Reviews. 2002;**2**:CD001520

[47] Kratz EM, Piwowar A, Zeman M, Stebelová K, Thalhammer T. Decreased melatonin levels and increased levels of advanced oxidation protein products in the seminal plasma are related to male infertility. Reproduction, Fertility, and Development. 2016;**28**:507-515

[48] Bonzini M, Palmer KT, Coggon D, Carugno M, Cromi A, Ferrario MM. Shift work and pregnancy outcomes: A systematic review with meta-analysis of currently available epidemiological studies. International Journal of Gynaecology and Obstetrics. 2011;**118**:1429-1437

[49] Stocker LJ, Macklon NS, Cheong YC, Bewley SJ. Influence of shift work on early reproductive outcomes: A systematic review and meta-analysis. Obstetrics and Gynecology. 2014;**124**:99-110

[50] Luboshizsky R, Lavie P. Sleep-inducing effects of exogenous melatonin administration. Sleep Medicine Reviews. 1998;**2**:191-202

[51] Hardeland R. Chronobiology of melatonin beyond the feedback to the

suprachiasmatic nucleus-consequences to melatonin dysfunction. International Journal of Molecular Sciences. 2013;**14**:5817-5841

[52] Cramer H, Rudolph J, Consbruch U, Kendel K. On the effects of melatonin on sleep and behavior in man. Advances in Biochemical Psychopharmacology. 1974;**11**:187-191

[53] Zhdanova IV, Wurtman RJ, Morabito C, Piotrovska VR, Lynch HJ. Effects of low oral doses of melatonin, given 2-4 hours before habitual bedtime, on sleep in normal young humans. Sleep. 1996;**19**:423-431

[54] Garfinkel D, Laudon M, Nof D, Zisapel N. Improvement of sleep quality in elderly people by controlled-release melatonin. Lancet. 1995;**346**(8974):541-544

[55] Hardeland R, Poeggeler B, Srinivasan V, Trakht I, Pandi-Perumal SR, et al. Melatonergic drugs in clinical practice. Arzneimittel-Forschung. 2008;**58**:1-10

[56] Pandi-Perumal SR, Srinivasan V, Maestroni GJ, Cardinali DP, Poeggeler B, et al. Melatonin: Nature's most versatile biological signal? The FEBS Journal. 2006;**273**:2813-2838

[57] Colten HR, Altevogt BM, editors. Sleep Disorders and Sleep Deprivation: An Unmet Public Health Problem. Washington, DC, US: National Academies Press; 2006

[58] Djokic G, Vojvodić P, Korcok D, Agic A, Rankovic A, Djordjevic V, et al. The effects of magnesium-melatonin-Vit B complex supplementation in treatment of insomnia. Open Access Macedonian Journal of Medical Sciences. 2019;7(18):3101-3105

[59] Abbasi B, Kimiagar M, Sadeqhniiat K, Shirazi MM, Hedavati M, Rashidkhani B. The effect of magnesium supplementation on primary insomnia in elderly: A double-blind placebo-controlled clinical trial. Journal of Research in Medical Sciences: The Official Journal of Isfahan University of Medical Sciences. 2012;**17**(12):1161-1169

[60] Syed EU, Wasay M, Awan S. Vitamin B12 supplementation in treating major depressive disorder: A randomized controlled trial. The Open Neurology Journal. 2013;**7**:44-48

[61] Williams AI, Cotter A, Sabina A, Girard C, Goodman J, Katz DL. The role of vitamin B6 as treatment for depression: A systematic review. Family Practice. 2005;**22**(5):532-537

[62] COMMISSION REGULATION (EU) No. 432/2012

Section 3

Antifibrilatory Potential of Melatonin

Chapter 4

Melatonin for a Healthy Heart Rhythm

Natalia Jorgelina Prado, Margarita Segovia-Roldan, Emiliano Raúl Diez and Esther Pueyo

Abstract

Melatonin is a promising cardioprotective agent. Its increase during the night is associated with healthy cardiovascular function. On the other hand, reduced levels of melatonin are related to diseases. Aging and chronodisruptors reduce melatonin levels. Pharmacological supplementation reduces the deleterious effects of cardiovascular risk factors and improves the myocardial response to ischemia/reperfusion injury and other proarrhythmic conditions. The protective mechanisms of melatonin involve its antioxidant properties as well as receptor-mediated actions. Signaling pathways include membrane responses, cytoplasmic modulation of kinases, nuclear receptor interactions, and improvement of mitochondrial functions. This chapter focuses on the electrophysiological and the antiarrhythmic properties of melatonin. The acute and chronic protective mechanisms of melatonin will be analyzed with an emphasis on transmembrane potentials and intercellular communication. An outstanding antifibrillatory effect makes melatonin a novel antiarrhythmic agent worthy of further exploration in the path to clinical applications.

Keywords: melatonin, arrhythmias, ventricular fibrillation, action potential, connexin 43, melatonin receptors

1. Introduction

"Nothing to do to save his life..." says the Beatles song "Good morning, good morning." Ironically, cardiovascular mortality and life-threatening arrhythmias show a circadian increase in the mornings, and chronoprotective agents are still missing [1, 2]. This chapter highlights the importance of melatonin as a potential life-saving agent for the darkest nights (of antiarrhythmics drugs) and a brightest tomorrow.

The cardioprotective properties of melatonin are remarkable. Most of the preclinical and clinical studies support the protective actions and the safety profile of this indolamine [3, 4]. In this chapter, we briefly introduce the multitarget and versatile properties of melatonin and general concepts of electrophysiology to appreciate its potential as a promising antiarrhythmic agent. The second and third sections of the chapters focus on acute and chronic melatonin's antiarrhythmic effects.

1.1 Melatonin properties relevant to heart rhythms

Endogenous and pharmacological increases of melatonin concentrations protect the cardiovascular system [3–11]. However, the relationships between the cardiovascular and circadian systems are highly complex and should not be interpreted in reductionist ways [5, 12–14]. Furthermore, our understanding of the pleiotropy of melatonin, a highly preserved molecule of protection, is continuously expanding [3–7, 10, 15–24]. Therefore, we will focus on melatonin effects on heart rhythms. Additional information regarding melatonin cardiovascular effects can be found elsewhere and include direct actions in the heart, blood vessels, kidney, and other regulatory mechanisms at the nervous, immune, and endocrine systems [11, 25, 26]. Only the electrophysiological information will be extracted from its protective actions against risk factors like hypertension, metabolic syndrome, obesity, inflammation, and pathologies like ischemia/reperfusion injury, infarction, drug-induced cardiotoxicity, diabetic cardiomyopathy, and heart failure [8, 11, 21, 27].

Melatonin is amphipathic and pleiotropic. Melatonin can act on several targets at cell membranes and at intracellular levels in almost any cell [28, 29]. For this electrophysiological analysis, we present the following division of melatonin mechanisms of action:

a. Antioxidant

b. Receptor activation

c. Improvement of mitochondrial functions

d. Ion channel modulation

1.1.1 Melatonin as an antioxidant

Melatonin protects against oxidants by several mechanisms. In fact, it has been suggested that one of the main functions of melatonin in all living organisms is to protect them from oxidative stress [30, 31]. Melatonin has a well-characterized and extensively documented antioxidant capacity [31–37]. Melatonin is a powerful antioxidant, with a potency of up to 10 times greater than vitamin E [38].

There are oxidants of different chemical nature. They can be free radicals or non-radical reactive species [39, 40]. Free radicals—molecules with an unpaired electron—are unstable, highly reactive, and often trigger chain reactions, which propagate nearby molecular modifications. The most studied oxidants are the reactive oxygen species (ROS), reactive nitrogen species (RNS), and reactive sulfur species. Under physiological conditions, ROS/RNS act as second intracellular messengers modulating signal transduction pathways [40, 41]. A delicate cellular balance between the production and the removal of free radicals maintains low/moderate concentrations. Oxidative stress occurs when oxidants increase above healthy levels and represent a severe risk to the molecular integrity of lipids, proteins, and DNA [39, 40]. Therefore, neutralization of reactive species by scavenger molecules like melatonin is a chemical way of counteracting oxidative stress.

The main agent involved in oxidative damage is superoxide anion, but hydrogen peroxide, hydroxyl radical, nitric oxide (NO), peroxynitrite, and nitroxyl also participate in oxidative stress. The mitochondria are the main source of oxidizing species during oxidative phosphorylation. Oxidants are also the product of the activation of non-mitochondrial enzyme systems such as NADPH oxidase, xanthine oxidase, and nitric oxide synthase [40–42].

Cells have antioxidants that prevent damage. An antioxidant is any substance that significantly delays or prevents oxidation of lipids, proteins, or DNA [40]. Lipids are often used as target molecules because they are more reactive to oxidants than proteins or DNA. Nonenzymatic antioxidants include reduced glutathione (GSH), vitamins, and melatonin among others. Melatonin is five times more effective than GSH as scavenger of the highly toxic hydroxyl radical [34]. The main antioxidant enzymes are superoxide dismutase (SOD), catalase, thioredoxin, and glutathione peroxidase [40–44].

Melatonin efficiently prevents oxidative stress. The aromatic indole ring of melatonin reduces and repairs electrophilic radicals acting as generous electron donor. One molecule of melatonin can neutralize up to 10 toxic reagents, including ROS, RNS, and other free radicals [7, 39, 45–47]. In addition, several metabolites formed when melatonin neutralizes harmful reagents are also antioxidants suggesting that a cascade of reactions increases the efficacy of melatonin [28, 35, 47–49]. Being a highly lipophilic and hydrophilic compound, melatonin crosses all morphological barriers and acts not only in each cell but also within each subcellular compartment. Additionally, melatonin increases the efficacy of vitamin E, vitamin C, and GSH [33, 50]. Therefore, the elimination of free radicals can be carried out by intracellular interactions independent of any receptor [36, 45, 51].

Melatonin stimulates antioxidant enzymes by acting on membrane, cytoplasmic, and nuclear receptors [39, 43, 52]. Low melatonin concentrations increase the expression or activity of SOD, catalase, and glutathione peroxidase [43, 53].

Ion channels and many other proteins respond to oxidative stress [54–58]. Amino acid residues are the targets of ROS/RNS. Sulfur atoms like cysteine and methionine, hydroxyl groups from tyrosine, or aromatic rings of histidine, phenylalanine, and tryptophan are vulnerable to reactive species. Those that contain more cysteines are more sensitive to changes because thiol groups (–SH), which exist as thiolates (–S) at physiological pH, tend to react more quickly with ROS/RNS [59]. Many of these proteins are involved in important biological reactions such as oxidative phosphorylation, metabolic regulation, and signal transduction [60, 61]. Oxidative stress can increase late sodium currents through direct Na^+ channel modification [62, 63] and result in a prolonged action potential duration and arrhythmogenic triggers known as early-after depolarizations (EAD) [64]. Several reviews describe the redox regulation of calcium channel in cardiac myocytes including the ryanodine receptor calcium, the IP3 receptor, and voltage-dependent L-type calcium channel [65–69]. ROS and RNS affect the L-type Ca^{2+} channel Cav1.2 by regulation of cysteine residues. However, calcium channel regulation by redox is controversial with reports of increase and decrease of channel functions [66]. Voltage-gated potassium (Kv) channel, mainly responsible for myocardial repolarization, is sensitive to oxidative stress [58, 70–72]. Sulfenic acid modification at a conserved cysteine residue of Kv1.5 under prolonged oxidative stress can induce arrhythmia [58, 72].

1.1.2 Melatonin receptors

Melatonin has receptors in the cellular membranes, in the cytoplasm, and in the nucleus. Melatonin membrane receptors express in several regions of the nervous system and in almost all the organs including the heart, arteries, kidneys, liver, gastrointestinal tract, prostate gland, uterus, skin, and eyes [73]. Melatonin activates two subtypes of G-protein-coupled receptors in the plasma membrane, named MT1 and MT2, according to the official IUPHAR nomenclature (previously called Mel1a and Mel1b) [74]. Both receptors have high affinity to melatonin ($K_d \sim 0.1$ pM). In 2019, Stauch and Johansson reported the crystal structures of the human MT1 and MT2 and set a solid base concerning ligand recognition for both receptors [75, 76].

Melatonin membrane receptors can exist as monomers, as well as dimers. The MT1 homodimer forms 3- to 4-fold higher proportion than the MT2 homodimer and the MT1/MT2 heterodimer. Nonmammalian vertebrates present a third low-affinity receptor termed Mel1c, and a proposed mammalian homologous is the orphan receptor GPR50 [74, 77–79]. This orphan lost its properties to directly interact with melatonin but shows an inhibitory interaction with MT1 receptors by forming heterodimers. More recently, other orphans unable to bind melatonin like GPR61, GPR62, and GPR135 showed a similar indirect inhibitory interaction with MT2 receptors [80]. Other G-protein-coupled receptors like the serotonin receptor 5HT2c can interact with melatonin membrane receptors [79]. These interesting interactions of membrane receptors are not further discussed in this chapter but should be considered in future electrophysiological studies with melatonin.

The MT1 and MT2 inhibit adenylate cyclase-protein kinase A-CREB signaling in target cells by pertussis toxin-sensitive Gαi, β, and γ and toxin-insensitive Gq, β, and γ proteins [74, 79]. The MT1 also increases phosphorylation of mitogen-activated protein kinase 1/2 (MAPK) and extracellular signal-regulated kinase 1/2 (ERK), as well as increasing potassium conductance through inwardly rectifying (Kir3.x) channels. The later effect on potassium channels could be relevant to heart electrophysiology since Kir3.x channels are highly expressed in cardiomyocytes and usually coupled to acetylcholine and adenosine membrane receptors [81]. MT2 melatonin receptor activation inhibits both forskolin-stimulated cAMP production and cGMP formation, activates protein kinase C (PKC) in the nervous system, and decreases calcium-dependent dopamine release in the retina. Native functional MT1/MT2 heterodimers in mouse rod photoreceptors mediate melatonin's enhancement of scotopic light sensitivity through phospholipase C and PKC pathways [82].

Several compounds interact with MT1 and MT2 receptors, but blocker luzindole is the only with proven myocardial electrophysiological effects [83]. Luzindole and 4P-PDOT competitively block MT1 melatonin receptors at concentrations higher than 300 nM, and both are inverse agonists in systems with constitutively active MT1 receptors [74, 79].

Melatonin interacts with several enzymes and intracellular proteins. The MT3 receptors is a quinone reductase 2 with an affinity in the nanomolar ranges [84]. This enzyme is possibly involved in the regulation of cellular oxidative status, although the exact regulatory action of melatonin remains unclear [84–87]. Furthermore, the electrophysiological effects of MT3 have not been reported yet.

Melatonin interacts with intracellular proteins such as calmodulin, calreticulin, or tubulin [88]. The low-affinity interaction between melatonin and calmodulin antagonizes the binding of Ca^{2+} and may be involved in its antioxidant action as well as other electrophysiological signaling processes [89–96].

Melatonin increases the cytoplasmic levels of the heat shock protein 70 in several tissues including the heart [97–102]. Further interaction with this chaperon will be described in Section 3 of the chapter.

Melatonin is a ligand for the retinoid-related orphan nuclear hormone receptor family (RZR/ROR) [74, 79]. RZR/RORα is expressed in a variety of organs, whereas RZRβ is specific for the brain and retina [33]. ROR/RZR has been proposed to work in coordination with the plasma membrane receptors MT1/MT2 to regulate gene expression. We suggest a potential interaction with Vitamin D receptor (VDR), which was elegantly confirmed in recent experiments [97, 103].

1.1.3 Melatonin improves mitochondrial functions

Mitochondria are critical for cellular metabolism and energy production. They maintain life but also are gatekeepers of cell death [31, 104]. Mitochondria produce up to 95% of the cellular energy in the form of ATP in aerobic cells [105]. Mitochondrial oxidative phosphorylation uses a system of oxidoreductase protein complexes (complexes I, II, III, and IV) to transfer electrons during ATP production. Deficiencies in the electron transport chain can result in the leakage of electrons and generate ROS/RNS [40, 41, 106, 107]. Oxidative stress decreases respiratory complex activity, impairs electron transport system, and opens the mitochondrial permeability transition pores leading to cell death [104, 106, 108].

Mitochondria are essentials for the protective actions of melatonin [51, 97, 106, 107, 109–111]. The mechanisms involved include its antioxidant properties and the preservation of complex I and III functions, inhibition of the opening of the permeability transition pores, and the release of cytochrome c. Petrosillo et al. demonstrate that melatonin prevents the opening of the mitochondrial permeability transition pores and its deleterious consequences [51, 110, 112, 113]. We recently reported that melatonin prevents mitochondrial edema, dilation of the ridges, high activity of NADPH oxidase, and apoptosis [97]. Melatonin improves mitofusin-2, which preserves the mitochondrial functional network and prevents apoptosis [114]. The reduction of mitochondrial damage in the heart could be related to the negative regulation of angiotensin II type 1 receptor (AT1) by melatonin [97]. The induction of Hsp70 through melatonin is compatible with an additional mechanism related to Tom 70, a translocase of the outer mitochondrial membrane [97, 115, 116]. The interaction of Hsp70 with Tom 70 initiates mitochondrial import processes [116]. Tom 70 regulates melatonin-induced cardioprotection by preventing mitochondrial deterioration and oxidative stress [97, 115].

Melatonin's cardioprotection associates with an increase in the number of mitochondria and positive regulation of survival genes such as nicotinamide phosphoribosyl transferase and nicotinamide adenine dinucleotide-dependent deacetylases, called sirtuins [117]. Particularly sirtuin-1 and sirtuin-3 are downstream mediators of the cardioprotective actions of melatonin. Sirtuin-1can modulate fatty acid oxidation, apoptosis, oxidative stress, and autophagy through deacetylation of transduction factors like NF-κB, forkhead box class O, p53, peroxisome proliferator-activated receptor alpha, thioredoxin-1, and Bcl-xL [117–121]. Sirtuin-3 is a family member that is primarily located in the mitochondria and protects against inflammation and diseases related to oxidative stress. Melatonin elevates sirtuin-3, stimulates superoxide dismutase activity, and suppresses mitochondrial oxidative stress [31, 117, 122, 123]. Additionally, melatonin protects nuclear and mitochondrial DNA [122, 124, 125]. The multiple actions of melatonin provide potent protection against mitochondrial-mediated lesions.

1.1.4 Melatonin modulates ion channels

Melatonin exerts its electrophysiological effects by multiple mechanisms. One of the ways for melatonin to interact is through the modulation of ion channels. Whether we consider its role as a drug or as a biological molecule, it should be taken into account how melatonin has been considered an electrophysiological modulator for many physiological and clinical conditions such as control of circadian rhythms, regulation of arterial blood pressure and heart rate in mammals, sleep processes, and antiaging, among others. Its role in the modulation of several ion channels is crucial to understand the molecular mechanism underlying the electrophysiological properties as an antiarrhythmic.

Melatonin regulates anionic and cationic selective channels by multiple pathways, at different doses and time-dependent responses. It is important to remember the wide spectrum of action this molecule has. For example, results regarding the pathophysiology of lung fibrosis show that volume-regulated anion currents do not respond to acute exposure of cells to melatonin in hypotonic solutions [126]. However, when cells are pre-incubated with melatonin concentrations from 1 to 100 μM for 30–60 min, the anionic currents in response to hypotonicity are blunted in a dose-dependent manner. These time- and dose-dependent responses could support the electrophysiological effect during regional ischemia after 20–30 minutes of melatonin exposure in isolated rat hearts, because during ischemia cardiomyocyte swelling activates anionic currents, and melatonin downregulation of these currents is a potential explanation [127, 128]. Additionally, these MT receptor interactions described in fibroblast deserve further evaluation in myocardial tissue.

From the perspective of the interaction between melatonin and its target, it will be crucial to increase the knowledge about the allosteric contact between melatonin and an ion channel. For example, melatonin blocks the potassium channels (Kv1.3) in a reversible manner through the interaction with different binding sites on the human peripheral blood T lymphocytes [129]. However, the inhibitory effects require high extracellular melatonin in the mM range [129]. Cardiomyocytes do not express this specific potassium channel, but a homologous mechanism can exist for other channels waiting to be reported.

Most of the information regarding the role and effect of melatonin in the organisms has been described in the nervous system. One of the most popular is melatonin-related circadian rhythm. In particular, how melatonin influences circadian phase and electrical activity thanks to the interaction with Kir3.x channels presents them as a therapeutic target for diseases related to circadian disruption and melatonin signaling features [130]. In addition, the effects of melatonin in this pacemaker of circadian rhythm could be due also to its modulation of inwardly rectifying potassium channels (Kir3.1/Kir3.2) via MT1 receptors [131]. Moreover, melatonin is also necessary for circadian regulation of sleep. This effect was described to be driven by the suppression of GABAergic neurons by melatonin in the lateral hypothalamus (crucial function for wakefulness), via interaction with MT1 receptor in order to inactivate hyperpolarization-activated cyclic nucleotide-gated channels [132].

Melatonin is a potential neuroprotective molecule thanks to its interaction in a mitochondrial pathway involving the closing of permeability transition pore and opening of ATP sensitive potassium channels (KATP) [133]. The opening of KATP contributes to melatonin antiseizure effect [134]. The preventive actions on the permeability transition pore have been reported in myocardial tissue as well [51, 112, 113]. However, opening of KATP channels with high concentrations of melatonin could be proarrhythmic [135, 136].

Melatonin modulates most of the voltage-activated calcium channel subtypes (L, P, Q, N, and R) with different effects [137–141]. Melatonin inhibits voltage-dependent calcium entry in cultured rat dorsal root ganglia neurons, regulates calcium entry into pineal cells, and has dose-dependent inhibitory effects on free $[Ca^{2+}]_i$ in mouse brain cells [137]. Melatonin has no effect on voltage-activated calcium channels in cultured human aortic smooth muscle cells [141]. Melatonin accutelly increase L type calcium currents in chick cardiac membranes [140, 141]. An early study shows that melatonin downregulates voltage-sensitive calcium channels in the heart [142]. These results indicate that melatonin may have differnt acute and chronic implications for normal cardiac physiology and for the pharmacological manipulation of the heart [142].

Melatonin mediates vasodilation of cerebral arteries through the activation of large-conductance Ca^{2+}-activated K^+ (BK_{Ca}) channels via both melatonin receptor-dependent and melatonin receptor-independent modes, increasing BK_{Ca} channel current density but not the KV channel current density [143]. Small-conductance

Ca^{2+}-activated K (SK) channels are also modulated by the action of melatonin [144]. Upregulation of SK channels plays a role in memory loss and indicates that melatonin reverses memory deficits in rats by downregulation of SK1, SK2, and SK3 channels in their hippocampi [144].

Additional information was brought about KCNQ from the aorta and related with vascular tone, and KCNH2 in the left ventricle was associated with QT duration in rats where melatonin was able to prevent the increase in blood pressure and change KCNQ and KCNH2 gene expression profiles [145].

Melatonin effects on connexin proteins will be extensively analyzed in the second and third sections of the chapters for its proven relationship with both acute and chronic antiarrhythmic effects of melatonin.

1.2 Electrophysiology and arrhythmias

The heart pumps blood under a synchronized electrical control. Arrhythmias are the electrical problems in the rhythm of the heart. The heartbeats may be faster in the case of tachyarrhythmia and slower in bradyarrhythmia.

Fatal arrhythmic events follow a circadian pattern [2]. Arrhythmogenesis decreases during nighttime when the melatonin levels increase 30 to 70 folds. Life-threatening cardiac arrhythmias (ventricular tachycardia, ventricular fibrillation, and sudden cardiac death) are more likely to occur in the morning after waking. Arrhythmias also increase with age and heart diseases [146–148].

Disturbances in membrane excitability or conduction cause arrhythmias. Excitability manifests as action potentials and involves coordinated ion movements across the cell membrane through ion channels, exchangers, and ATPases [149]. Conduction is the propagation of bioelectrical signal throughout the heart. Action potentials automatically originate at the sinoatrial node, spread to the atria, and, after a small delay in the atrioventricular node, rapidly and synchronously activate the ventricles via the His-Purkinje system. Action potentials propagate from cell to cell using low-resistance pathways known as gap junctions. Connexin proteins assemble into intercellular channels at gap junctions. Connexin 43 (Cx43) is the most abundant connexin in the heart [150]. Gap junctions couple the cells and allow the flow of electrical current and small molecules. The largest accumulation of connexins occurs in a specialized structure at the ends of cardiomyocytes called intercalated discs. Cardiac propagation is anisotropic, particularly more rapid in the longitudinal direction of the cell than in the transverse direction. The lateral borders of the myocytes usually show variable amount gap junctions depending on age or disease.

Cardiovascular diseases are the leading cause of death in the world [151]. Most deaths occur suddenly [152]. Catastrophic sudden death events motivate us to search for causes and possible solutions [153]. This is a great scientific and social health challenge. The approaches of recent years have reduced the burden of cardiovascular disease, but there is still much to improve [154]. A case occurs with arrhythmias. The rhythm disorders motivated emergency interventions, especially during the first hour of the manifestation of coronary heart disease. Cardiopulmonary resuscitation, ambulances, and cardiodefibrillation were response strategies to unexpected events. Unfortunately, they are still unexpected due to the limited understanding of the causes at a level that would allow us to predict, avoid, or control the occurrence of an event [155]. In that sense, the strategies that attempt to determine risks grew in order to establish a more efficient direction of interventions [156, 157]. Today they allow us to expect more lethal events in severely ill people. However, risk factors are still far from being effective and much less efficient. The changes that occur in physiology as a result of exposure to different risk factors would be one of the explanations [158].

2. Acute antiarrhythmic mechanisms of melatonin

Melatonin acts at multiple electrophysiological levels due to its receptor-dependent and receptor-independent mechanisms. In 1998, the seminal work of Tan et al. highlighted the antiarrhythmic properties of melatonin [159]. During the past two decades, our understanding of the pleiotropic action of melatonin increased significantly.

The antiarrhythmic effect of melatonin was first attributed to its notable antioxidant properties, mainly because melatonin results were better than those obtained with an ascorbic acid at concentration 10–500 times higher [159]. Numerous studies confirmed antiarrhythmic protection and related it to its remarkable antioxidant properties [160–167].

Our research group corroborated the antiarrhythmic effect of melatonin in isolated hearts of female rats, when administered continuously from the stage prior to the onset of myocardial ischemia [127]. Notably, the antiarrhythmic protection had a dose-dependent response, while the antioxidant capacity was the same for all the doses studied. The preventive effect on the shortening of the action potential that occurs between the 7th and 10th minute of the ischemia was another dose-dependent variable found in our study. This led us to think that the antiarrhythmic mechanism could be due to a lower heterogeneity in the repolarization of myocardial tissue that diminishes the possibility of reentry circuits being formed and maintained. As previously mentioned, the time- and dose-dependent responses could be due to melatonin inhibitory effect against swell-activated anionic currents [126–128].

We recently showed that melatonin reduces arrhythmias when administered during reperfusion, a useful timing for the clinical context of acute coronary syndromes, because most therapies can only start close to the reperfusion period [168]. Melatonin showed protective mechanisms when administered to isolated hearts of rats fed with fructose and spontaneously hypertensive rats. These animals show greater activity of the enzyme NADPH oxidase, which is one of the main systems for generating free radicals, and, therefore, higher levels of oxidative stress. The antiarrhythmic effect was not affected in the models with greater oxidative stress, and in all groups, it was accompanied by a temporary shortening of the duration of the action potential during the first 3–5 minutes of reperfusion. This result was interpreted as a reduction in the ability to generate early and late postdepolarizations. Self-limited arrhythmic events, such as ventricular extrasystoles, salvos, and even non-sustained ventricular tachycardia, occurred in all experimental groups. The main difference was that the hearts treated with melatonin did not show sustained forms of arrhythmias, either sustained ventricular tachycardia or ventricular fibrillation. These results (potential shortening and absence of sustained arrhythmia) are difficult to reconcile with the mechanisms postulated for reentry circuits.

The same year of our publication of the antiarrhythmic protection of melatonin administered in reperfusion, another group published that melatonin protects against arrhythmias, by increasing the threshold to electrically induce sustained ventricular fibrillation, by increasing the myocardial Cx43 by PKC in hypertensive rats [169]. Melatonin prevented myocardial abnormalities of connexin and improved cardiac conduction.

Based in these interesting results, we tested if melatonin could prevent hypokalemia-induced ventricular fibrillation by Cx43 preservation [83]. The acute administration of melatonin during low potassium perfusion reduced the incidence of ventricular fibrillation and improved the recovery of sinus rhythm in those hearts that, despite being treated with melatonin, developed sustained fibrillation. Protection was mediated by the activation of melatonin receptors and by the prevention of dephosphorylation and lateralization of Cx43.

A brief explanation of the electrophysiological changes induced by hypokalemia will help to appreciate the relevance as antiarrhythmic. Severe hypokalemia induces changes in ventricular repolarization, such as lengthening the QT interval, prominent U waves, fusion of T and U waves associated with and increases risk of arrhythmic death [83, 170, 171]. Our experimental model confirmed the lengthening of the QT interval and correlated with an increase in the duration of the action potential [83]. Melatonin did not prevent the prolongation of the action potential induced by hypokalemia when measured at 90% of repolarization but maintained action potential duration at 50% of repolarization and made the membrane potential more stable, showing less after depolarization. Luzindole blunted both effects of melatonin, suggesting the involvement of melatonin receptor activation in the preservation of membrane potential.

Hypokalemia decreases NaK-ATPase activity and causes an intracellular Ca^{2+} overload that facilitates the development of delayed postdepolarizations through the transient inward currents [172–174]. Delayed postdepolarizations are considered triggers of arrhythmias because they can initiate an action potential in isolated cells. However, it is unlikely that an extra action potential can be initiated from a single cell in the tissue due to a mismatch between the current source from the cell and the current sink produced by the surrounding cells [175]. To overcome the source-sink mismatch, there must be a reduced sink through intercellular decoupling or an increase in the source through the synchronization of delayed postdepolarizations between several adjacent cells. Both situations could be assumed based on the results of anisotropic conduction studies and immunofluorescence imaging [83].

In fact, hypokalemia induces conduction abnormalities, increased amplitude and duration of the P wave, a slight prolongation of the PR interval, atrioventricular block, increased QRS duration, and cardiac arrest [173, 176]. We found all these electrocardiographic disorders during our experimental model of hypokalemia [83]. Melatonin prevented the widening of the QRS and delayed activation of the potential for epicardial action. The latter could be considered as a substitute for conduction velocity in complex tissues such as ventricles, assuming unknown routes from endocardial activation points that indicate the onset of QRS to epicardial myocytes recorded with microelectrodes. These improvements in ventricular conduction were related to Cx43 lateralization and dephosphorylation.

The lateralization of connexins has been detected in chronic atrial fibrillation, cardiac hypertrophy, heart failure, and after myocardial infarction [21, 177–179]. An increase in the fraction of lateral connexins that form functional channels improves transverse conduction velocity and contributes to the spread of the arrhythmogenic impulse. High side-by-side lateralization can favor conduction blockage due to mismatches between the source and the sink [175, 180]. A unidirectional block can lead to reentry circles that result in tachycardia or ventricular fibrillation [181]. Therefore, the acute lateralization induced by hypokalemia is an important arrhythmogenic factor [83]. It is noteworthy that melatonin prevented acute lateralization of Cx43.

Connexin 43 phosphorylation could lead to better coupling or uncoupling depending on the target amino acid, but dephosphorylation is clearly associated with uncoupling [21, 177, 182]. It is not yet known whether the dephosphorylation of Cx43 during low potassium is the result of increased phosphatase activity and/or an increase in phosphokinase or what are the intracellular mechanisms that prevented dephosphorylation when treated with melatonin. Dramatic reductions in intercellular communication due to the loss of phosphorylated Cx43 and the accumulation of nonphosphorylated Cx43 were previously reported in other experimental models [177].

Our results could be relevant mainly in those situations in which acute hypokalemia can be anticipated as in dialysis [183, 184]. Both QT interval and the QT dispersion increase after dialysis. We propose that melatonin could make the heart

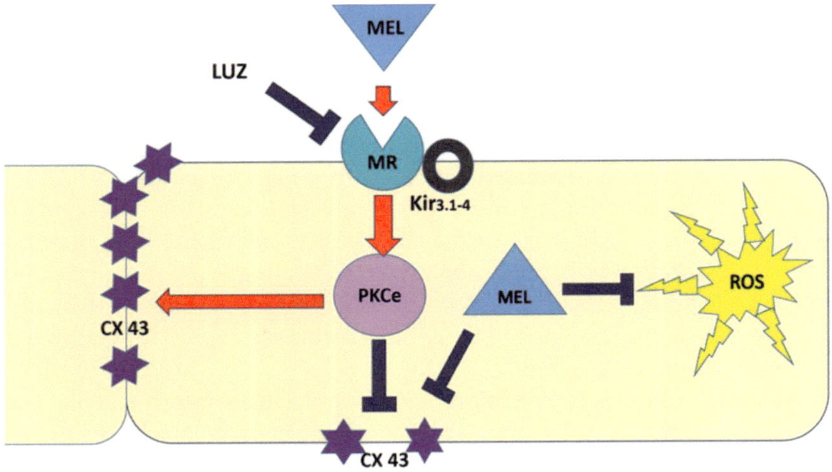

Figure 1.
Acute antiarrhythmic mechanisms of melatonin. The red arrows indicate stimulation, and the interrupted blue lines indicate blockage.

more resistant to arrhythmic events triggered by rapid changes in plasma electrolyte concentrations, regardless of a lack of effects on the ECG. In addition, those dialysis patients also suffer from disorders in the circadian rhythms and low levels of melatonin [185]. However, clinical translations of our results should be done with caution, mainly because we use a high dose of melatonin administered directly to the heart. Based on melatonin's pharmacokinetics in humans, to achieve a similar concentration in plasma to the one tested *ex vivo*, a dose 10 times higher the highest intravenous dose tested until now should be administered [186, 187].

Melatonin has a remarkable antiarrhythmic activity that is carried out based on actions dependent on and independent of receptor activation. To summarize we propose that the antiarrhythmic effect of melatonin is mediated by receptor activation beyond its outstanding antioxidant actions (**Figure 1**). The shortening of the action potential could be associated with the activation of MT1 melatonin receptors, since they can regulate specific ion channels such as Kir3.1 channel. MT1 and MT2 receptors could indirectly modulate other electrophysiological effects through intracellular signaling such as decreased cyclic adenosine monophosphate, increased phospholipase C, and PKC activation.

3. Chronic antiarrhythmic mechanisms of melatonin

Endogenous melatonin would be an intrinsically protective factor with therapeutic potential [188, 189]. Melatonin is a promising treatment for cardiovascular diseases such as myocardial ischemia/reperfusion injury, hypertension, and heart failure. It has been shown that melatonin levels were reduced in patients with acute myocardial infarction and in patients undergoing primary coronary angioplasty [190]. These findings suggest that melatonin could play an important role in preventing ischemia/reperfusion heart injury. Indeed, reperfusion arrhythmias increase in pinealectomized animals, suggesting a protective role of endogenous physiological melatonin levels [163, 189].

Chronic melatonin supplementation, either in physiological or pharmacological ranges, protects against arrhythmias [8, 21, 97, 163, 169, 189, 191, 192]. Beyond the reported antioxidant properties of melatonin, it reduces severe ventricular arrhythmias by antifibrotic mechanisms, electrical remodeling, direct mitochondrial protection, myocardial Cx43 preservation via PKC signaling, and vitamin D-HSP70/AT1 counterbalance (**Figure 2**). Its cardioprotective properties persist in relevant cardiovascular risk factor models like hypertensive, obese, and nephropathic rats. The latter is interesting because most of the therapeutic interventions postulated so far fail to be reproduced under risk factor conditions.

A preventive approach would be of great value in the face of unpredictable acute arrhythmic events, especially if the intervention manages to avoid the most severe and potentially lethal arrhythmias such as ventricular tachycardia and fibrillation. Numerous efforts have been made in that direction. In the last quarter of the twentieth century, several antiarrhythmic drugs were tried, but most of them showed a proarrhythmogenic profile or failed to reduce mortality [193–196]. A time of great progress was appreciated with the introduction of implantable cardiodefibrillators. However, surgical intervention and high cost limit its population efficiency. A strategy to improve the availability of preventive interventions is to select potential beneficiaries based on their risk of serious events. This would compensate for potential side effects and optimize the investment of resources. Other strategies, such as vaccines, are based on achieving the greatest possible scope with the least number of interventions that attenuate the severity of diseases. In the case of arrhythmias, we still have no clear "antiarrhythmic vaccine." Therefore, risk-oriented strategies would be an acceptable approach.

From a preventive point of view, the pleiotropic protection mechanisms of melatonin could effectively limit the arrhythmic complications associated with hypertension [21, 169]. Arterial hypertension causes vascular deterioration, overloads the heart, and predisposes to a greater number of arrhythmic events. More than five decades ago, it was reported that surgical removal of the pineal gland, a procedure that essentially eliminates circulating levels of melatonin, was followed by a slow but persistent increase in blood pressure in rats [197]. This finding has been confirmed in several subsequent studies [11]. In addition, daily treatment of pinealectomized rats with melatonin attenuated the elevation in blood pressure

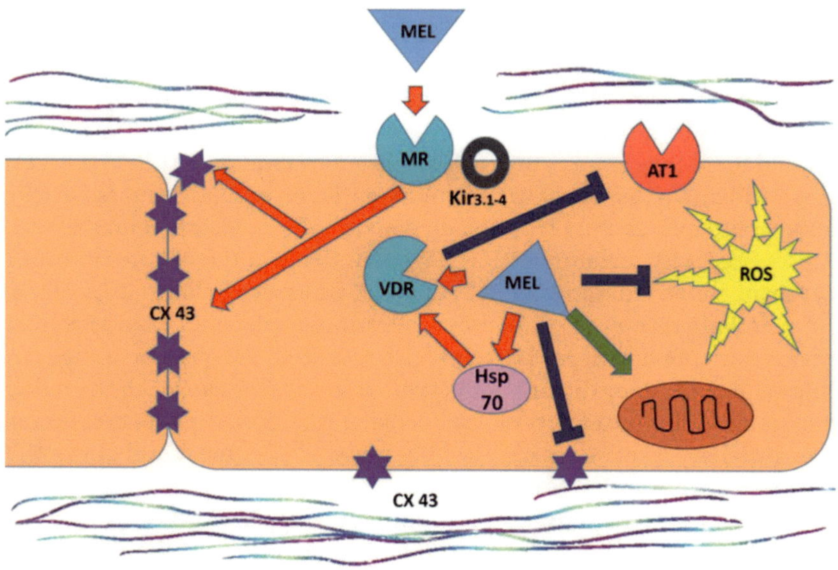

Figure 2.
Antiarrhythmic mechanisms of chronic melatonin administration. Extracellular lines represent reduced fibrosis after melatonin treatment. The red arrows indicate stimulation and blue ones show blockage. The green arrow marks direct mitochondrial protection.

that accompanies pinealectomy [198, 199]. Potentially related to these experimental findings are those observational studies in humans that document an age-related gradual increase in blood pressure [7]. Of special interest is that the ability of the pineal gland to produce melatonin is compromised during aging so that the levels of melatonin in the blood at night gradually decrease [28, 30, 110]. An implication of these findings is that the loss of melatonin during aging can contribute to gradual hypertension and arrhythmias.

The structural remodeling of the myocardium that follows hypertension (mainly cardiomyocyte hypertrophy and fibrosis) is accompanied by changes in the expression, distribution and function of the ionic channels of the cell membrane and the intercellular channels constituted by Cx43 [21, 191, 200]. Remodeling predisposes to life-threatening ventricular tachycardia and ventricular fibrillation by early or late postdepolarization and reentry. Melatonin prevents changes in ventricular redistribution of Cx43 and reduces arrhythmia inducibility [8, 21, 147, 191].

Another chronodisruptor that increases arrhythmic risk are kidney diseases. Chronic kidney diseases (CKD) alter the nocturnal secretion of melatonin [185, 201]. Melatonin levels correlate negatively with the intrarenal activity of renin-angiotensin II-aldosterone system (RAAS) [202]. Melatonin improves intrarenal RAAS in the 5/6 nephrectomy rat model and reduces blood pressure, oxidative stress, and interstitial fibrosis in the remaining kidneys [203].

Renal diseases cause cardiovascular and electrolytic remodeling that increases the risk of arrhythmias [204–206]. Cardiovascular events occur more frequently in patients with chronic kidney disease. Ventricular arrhythmias are particularly prevalent among patients with CKD, even when those patients do not suffer from any electrolyte imbalance [207]. The risk of mortality also increases in patients with CKD who suffer from an acute coronary syndrome [208]. We demonstrated that unilateral ureteral obstruction caused a cardiac remodeling that was accompanied by an increase in reperfusion arrhythmias [209].

The electrophysiological properties of chronic melatonin deserve attention, due to their relevance for cardiorenal situations with high arrhythmic risk and lack of treatments. We recently confirmed the antifibrotic, antiapoptotic, and antioxidant effects of melatonin and linked them to an HSP 70-VDR/AT1 counterbalance which prevents kidney damage and arrhythmogenic remodeling of the heart [97].

In renal and myocardial tissue, melatonin increased HSP 70 and VDR and decreased AT1 and fibrosis. Melatonin increases HSP 70 and protects the liver of rats exposed to toluene from cytotoxicity induced by oxidative stress [100]. HSP 70 regulates antioxidant responses to cellular oxidative stress and reduces NADPH oxidase activity and expression [210]. We demonstrated a myocardial increase in HSP 70 in rats treated with melatonin. HSP 70 induces VDR and facilitates intracellular localization of active vitamin D metabolites and transactive VDRs [209, 211, 212]. Nuclear melatonin receptors, as members of retinoid-related orphan receptors, may interact and prevent degradation of VDR [97, 103]. Expression of myocardial VDR links chronic kidney disease with cardiovascular disease due to the reduction in VDR that amplifies the effects of angiotensin [212]. Melatonin decreases renal and myocardial overexpression of AT1 [97]. It is well documented that the AT1 pathway leads to myocardial fibrosis during CKD [97]. As previously suggested, the low expression of AT1 through VDR induction could be a consequence of HSP 70-mediated cellular protection [213]. Angiotensin II exerts a tonic modulation of melatonin synthesis by influencing the activity of tryptophan hydroxylase through AT1 supporting the postulated feedback (or reciprocal regulation) between AT1 and melatonin [97, 202].

Additionally, the mitochondrial dynamics relates to the RAAS. We show that melatonin prevents mitochondrial edema, high activity of NADPH oxidase, and apoptosis. In this sense, the reduction of mitochondrial damage melatonin could be related to the negative regulation of AT1. The induction of HSP 70 through melatonin is compatible with an additional mechanism related to Tom 70. Furthermore, Tom 70 regulates melatonin-induced protection against myocardial infarction [115, 116]. All these data allow us to assume that the induction of HSP 70 by melatonin and the reduction of AT1 are critical components of the cellular stress response.

We attribute the higher vulnerability to ventricular fibrillation during reperfusion in the kidney disease rat model to the prooxidative and profibrotic changes that accompanied the increase in AT1 and the decrease in HSP 70 [97, 209]. Myocardial oxidative stress—particularly in the mitochondria—and fibrosis are well-known proarrhythmic substrates [55, 71]. Free radicals act as triggers for the beginning of arrhythmic events. The persistence of high-frequency rhythms requires reentry circuits [214]. Altered conduction and shortening of the action potential contribute to the complex reentry mechanisms involved in ventricular fibrillation.

Melatonin protection against myocardial remodeling induced by kidney disease is one of the factors that protect against ventricular fibrillation. Chronic melatonin prolongs the action potential duration and hyperpolarizes the cardiomyocytes. These changes are the first report of myocardial action potential modifications by chronic administration of melatonin. Opening of Kir3.x channels by melatonin receptor activation could explain hyperpolarization [131]. The action potential lengthening is harder to explain because melatonin activates currents involved in the action potential repolarization and the only inhibitory effect of melatonin against outward potassium currents was described in neurons [90, 129, 130, 215–217]. As previously mentioned, the downregulation of volume-activated anionic currents can explain attenuated response to action potential shortening induced by ischemia [126].

A synthesis of the mechanisms of protection of chronic treatment of melatonin cardiovascular complications is outlined in **Figure 2**. We focus our attention on the preventive effects of melatonin against the alteration of Cx43, mitochondrial oxidant capacity, and membrane potentials. In addition, modulation of the AT1 and VDR receptors related to the increase of HSP 70 contributes to the cardioprotective effects of melatonin.

4. Conclusions

Melatonin is the rhythmic protector of healthy heart rhythm and a promising preventive agent against ventricular fibrillation, the most lethal and disorganized heart rhythm. Pleotropic effects of melatonin make it an exceptional acute and chronic antiarrhythmic.

Acknowledgements

This work was supported by grant by ESC research grant funded by the European Society of Cardiology, by ERC-2014-StG 638284 funded by the European Research Council, by project DPI2016-75458-R funded by MINECO (Spain) and FEDER, and by Reference Group BSICoS T39-17R and project LMP124-18 funded by Gobierno de Aragón and FEDER 2014-2020 "Building Europe from Aragón."

Conflict of interest

The authors declare no conflict of interest.

Author details

Natalia Jorgelina Prado[1,2], Margarita Segovia-Roldan[2], Emiliano Raúl Diez[1,2*] and Esther Pueyo[1,2]

1 Medical Faculty, CONICET, IMBECU, National University of Cuyo, Mendoza, Argentina

2 I3A, Universidad de Zaragoza, IIS Aragón and CIBER-BBN, Zaragoza, Spain

*Address all correspondence to: diez.emiliano@fcm.uncu.edu.ar

References

[1] Priori SG, Blomström-Lundqvist C, Mazzanti A, Blom N, Borggrefe M, Camm J, et al. 2015 ESC guidelines for the management of patients with ventricular arrhythmias and the prevention of sudden cardiac death. European Heart Journal. 2015;**36**:2793-2867

[2] Black N, D'Souza A, Wang Y, Piggins H, Dobrzynski H, Morris G, et al. Circadian rhythm of cardiac electrophysiology, arrhythmogenesis, and the underlying mechanisms. Heart Rhythm. 2019;**16**:298-307

[3] Baltatu OC, Senar S, Campos LA, Cipolla-Neto J. Cardioprotective melatonin: Translating from proof-of-concept studies to therapeutic use. International Journal of Molecular Sciences. 2019;**20**:4342

[4] Zhou H, Ma Q, Zhu P, Ren J, Reiter RJ, Chen Y. Protective role of melatonin in cardiac ischemia-reperfusion injury: From pathogenesis to targeted therapy. Journal of Pineal Research. 2018;**64**:e12471

[5] PlissMG, KuzmenkoNV, RubanovaNS, Tsyrlin VA. Dose-dependent mechanisms of melatonin on the functioning of the cardiovascular system and on the behavior of normotensive rats of different ages. Advances in Gerontology. 2019;**9**:327-335

[6] Jiki Z, Lecour S, Nduhirabandi F. Cardiovascular benefits of dietary melatonin: A myth or a reality? Frontiers in Physiology. 2018;**9**:528

[7] Favero G, Franceschetti L, Buffoli B, Moghadasian MH, Reiter RJ, Rodella LF, et al. Melatonin: Protection against age-related cardiac pathology. Ageing Research Reviews. 2017;**35**:336-349

[8] Egan Benova T, Viczenczova C, Szeiffova Bacova B, Knezl V, Dosenko V, Rauchova H, et al. Obesity-associated alterations in cardiac connexin-43 and PKC signaling are attenuated by melatonin and omega-3 fatty acids in female rats. Molecular and Cellular Biochemistry. 2019;**454**:191-202

[9] Singhanat K, Apaijai N, Chattipakorn SC, Chattipakorn N. Roles of melatonin and its receptors in cardiac ischemia–reperfusion injury. Cellular and Molecular Life Sciences. 2018;**75**:4125-4149

[10] Pandi-Perumal SR, BaHammam AS, Ojike NI, Akinseye OA, Kendzerska T, Buttoo K, et al. Melatonin and human cardiovascular disease. Journal of Cardiovascular Pharmacology and Therapeutics. 2017;**22**:122-132

[11] Prado NJ, Ferder L, Manucha W, Diez ER. Anti-inflammatory effects of melatonin in obesity and hypertension. Current Hypertension Reports. 2018;**20**:45

[12] Crnko S, Du Pré BC, Sluijter JPG, Van Laake LW. Circadian rhythms and the molecular clock in cardiovascular biology and disease. Nature Reviews. Cardiology. 2019;**16**:437-447

[13] Rana S, Prabhu SD, Young ME. Chronobiological influence over cardiovascular function. Circulation Research. 2020;**126**:258-279

[14] Martino TA, Young ME. Influence of the cardiomyocyte circadian clock on cardiac physiology and pathophysiology. Journal of Biological Rhythms. 2015;**30**: 183-205

[15] Zhou H, Yue Y, Wang J, Ma Q, Chen Y. Melatonin therapy for diabetic cardiomyopathy: A mechanism involving Syk-mitochondrial complex I-SERCA pathway. Cellular Signalling. 2018;**47**:88-100

[16] Yang Y, Sun Y, Yi W, Li Y, Fan C, Xin Z, et al. A review of melatonin as a

suitable antioxidant against myocardial ischemia-reperfusion injury and clinical heart diseases. Journal of Pineal Research. 2014;**57**:357-366

[17] Sun H, Gusdon AM, Qu S. Effects of melatonin on cardiovascular diseases. Current Opinion in Lipidology. 2016;**27**:408-413

[18] Han D, Wang Y, Chen J, Zhang J, Yu P, Zhang R, et al. Activation of melatonin receptor 2 but not melatonin receptor 1 mediates melatonin-conferred cardioprotection against myocardial ischemia/reperfusion injury. Journal of Pineal Research. 2019;**67**:e12571

[19] Hsu C-N, Huang L-T, Tain Y-L. Perinatal use of melatonin for offspring health: Focus on cardiovascular and neurological diseases. International Journal of Molecular Sciences. 2019;**20**:5681

[20] Li K, Hu F, Xiong W, Wei Q, Liu FF. Network-based transcriptomic analysis reveals novel melatonin-sensitive genes in cardiovascular system. Endocrine. 2019;**64**:414-419

[21] Egan Benova T, Szeiffova Bacova B, Viczenczova C, Diez E, Barancik M, Tribulova N. Protection of cardiac cell-to-cell coupling attenuate myocardial remodeling and proarrhythmia induced by hypertension. Physiological Research. 2016;**65**(Suppl 1):S29-S42

[22] Zhai M, Li B, Duan W, Jing L, Zhang B, Zhang M, et al. Melatonin ameliorates myocardial ischemia reperfusion injury through SIRT3-dependent regulation of oxidative stress and apoptosis. Journal of Pineal Research. 2017;**63**:e12419

[23] Nduhirabandi F, Maarman G, Nduhirabandi F, Maarman GJ. Melatonin in heart failure: A promising therapeutic strategy? Molecules. 2018;**23**:1819

[24] Reiter RJ, Tan DX, Paredes SD, Fuentes-Broto L. Beneficial effects of melatonin in cardiovascular disease. Annals of Medicine. 2010;**42**:279-285

[25] Pechanova O, Paulis L, Simko F. Peripheral and central effects of melatonin on blood pressure regulation. International Journal of Molecular Sciences. 2014;**15**:1792017937

[26] Baltatu OC, Amaral FG, Campos LA, Cipolla-Neto J. Melatonin, mitochondria and hypertension. Cellular and Molecular Life Sciences. 2017;**74**:3955-3964

[27] Lahera V, de Las HN, López-Farré A, Manucha W, Ferder L. Role of mitochondrial dysfunction in hypertension and obesity. Current Hypertension Reports. 2017;**19**:11

[28] Pandi-Perumal SR. Melatonin: Biological Basis of its Function in Health and Disease. CRC Press; 2005. ISBN 1-58706-244-5

[29] Hardeland R, Cardinali DP, Srinivasan V, Spence DW, Brown GM, Pandi-PerumalSR.Melatonin-apleiotropic, orchestrating regulator molecule. Progress in Neurobiology. 2011;**93**:350-384

[30] Poeggeler B, Reiter RJ, Tan D-X, Chen L-D, Manchester LC. Melatonin, hydroxyl radical-mediated oxidative damage, and aging: A hypothesis. Journal of Pineal Research. 1993;**14**:151-168

[31] Reiter RJ, Rosales-Corral S, Tan DX, Jou MJ, Galano A, Xu B. Melatonin as a mitochondria-targeted antioxidant: One of evolution's best ideas. Cellular and Molecular Life Sciences. 2017;**74**:3863-3881

[32] Galano A, Tan DX, Reiter RJ. Melatonin as a natural ally against oxidative stress: A physicochemical examination. Journal of Pineal Research. 2011;**51**:1-16

[33] Galano A, Reiter RJ. Melatonin and its metabolites vs oxidative stress: From individual actions to collective protection. Journal of Pineal Research. 2018;**65**:e12514

[34] Galano A. On the direct scavenging activity of melatonin towards hydroxyl and a series of peroxyl radicals. Physical Chemistry Chemical Physics (PCCP). 2011;**13**:7178-7188

[35] Galano A, Tan DX, Reiter RJ. On the free radical scavenging activities of melatonin's metabolites, AFMK and AMK. Journal of Pineal Research. 2013;**54**:245-257

[36] Manchester LC, Coto-Montes A, Boga JA, Andersen LPH, Zhou Z, Galano A, et al. Melatonin: An ancient molecule that makes oxygen metabolically tolerable. Journal of Pineal Research. 2015;**59**:403-419

[37] Reiter RJ, Tan DX, Galano A. Melatonin: Exceeding expectations. Physiology. 2014;**9**:325-333

[38] Reiter RJ, Tan DX. Melatonin: A novel protective agent against oxidative injury of the ischemic/reperfused heart. Cardiovascular Research. 2003;**58**:10-19

[39] Reiter RJ, Mayo JC, Tan D-X, Sainz RM, Alatorre-Jimenez M, Qin L. Melatonin as an antioxidant: Under promises but over delivers. Journal of Pineal Research. 2016;**61**:253-278

[40] Forman HJ. Redox signaling: An evolution from free radicals to aging. Free Radical Biology & Medicine. 2016;**97**:398-407

[41] Dröge W. Free radicals in the physiological control of cell function. Physiological Reviews. 2002;**82**:47-95

[42] Radi R. Oxygen radicals, nitric oxide, and peroxynitrite: Redox pathways in molecular medicine. Proceedings of the National Academy of Sciences of the United States of America. 2018;**115**:5839-5848

[43] Rodriguez C, Mayo JC, Sainz RM, Antolin I, Herrera F, Martin V, et al. Regulation of antioxidant enzymes: A significant role for melatonin. Journal of Pineal Research. 2004;**36**:1-9

[44] Afanas'Ev I. ROS and RNS signaling in heart disorders: Could antioxidant treatment be successful? Oxidative Medicine and Cellular Longevity. 2011;**2011**:1-13

[45] Yang Y, Sun Y, Yi W, Li Y, Fan C, Xin Z, et al. A review of melatonin as a suitable antioxidant against myocardial ischemia-reperfusion injury and clinical heart diseases. Journal of Pineal Research. 2014;**57**:357-366

[46] Brazão V, Santello FH, Colato RP, Mazotti TT, Tazinafo LF, MPA T, et al. Melatonin: Antioxidant and modulatory properties in age-related changes during *Trypanosoma cruzi* infection. Journal of Pineal Research. 2017;**63**:e12409

[47] Tan D-X, Manchester LC, Terron MP, Flores LJ, Reiter RJ. One molecule, many derivatives: A never-ending interaction of melatonin with reactive oxygen and nitrogen species? Journal of Pineal Research. 2007;**42**:28-42

[48] Tan DX, Hardeland R, Manchester LC, Paredes SD, Korkmaz A, Sainz RM, et al. The changing biological roles of melatonin during evolution: From an antioxidant to signals of darkness, sexual selection and fitness. Biological Reviews. 2010;**85**:607-623

[49] Tengattini S, Reiter RJ, Tan DX, Terron MP, Rodella LF, Rezzani R. Cardiovascular diseases: Protective effects of melatonin. Journal of Pineal Research. 2008;**44**:16-25

[50] Dominguez-Rodriguez A, Abreu-Gonzalez P, Garcia-Saiz MM, Aldea-Perona A, de la Torre JM, Garcia-Camarero T, et al. Cardio-protection with melatonin in the acute myocardial infarction: Awaiting results of MARIA trial? International Journal of Cardiology. 2015;**182**:54-55

[51] Petrosillo G, Colantuono G, Moro N, Ruggiero FM, Tiravanti E, Di Venosa N, et al. Melatonin protects against heart ischemia-reperfusion injury by inhibiting mitochondrial permeability transition pore opening. The American Journal of Physiology-Heart and Circulatory Physiology. 2009;**297**:H1487-H1493

[52] Yeung HM, Hung MW, Lau CF, Fung ML. Cardioprotective effects of melatonin against myocardial injuries induced by chronic intermittent hypoxia in rats. Journal of Pineal Research 2015;**58**:12-25

[53] Yanar K, Simsek B, Çakatay U. Integration of melatonin related redox homeostasis, aging, and circadian rhythm. Rejuvenation Research. 2019;**22**:409-419

[54] Sovari AA. Cellular and molecular mechanisms of arrhythmia by oxidative stress. Cardiology Research and Practice. 2016;**2016**:1-7

[55] Yang KC, Kyle JW, Makielski JC, Dudley SC. Mechanisms of sudden cardiac death: Oxidants and metabolism. Circulation Research. 2015;**116**:1937-1955

[56] Zhou L, Cortassa S, Wei AC, Aon MA, Winslow RL, O'Rourke B. Modeling cardiac action potential shortening driven by oxidative stress-induced mitochondrial oscillations in guinea pig cardiomyocytes. Biophysical Journal. 2009;**97**:1843-1852

[57] Akbarali HI. Oxidative stress and ion channels. In Systems Biology of Free Radicals and Antioxidants; Springer-Verlag Berlin Heidelberg, 2012; pp. 355-373; ISBN 9783642300189

[58] Bhuyan R, Chakraborti S. Oxidative stress and modulation of cardiac Kv1.5 channel. In: Oxidative Stress in Heart Diseases. Singapore: Springer; 2019. pp. 191-203

[59] Poole LB. The basics of thiols and cysteines in redox biology and chemistry. Free Radical Biology & Medicine. 2015;**80**:148-157

[60] Chouchani ET, James AM, Methner C, Pell VR, Prime TA, Erickson BK, et al. Identification and quantification of protein S -nitrosation by nitrite in the mouse heart during ischemia. The Journal of Biological Chemistry. 2017;**292**:14486-14495

[61] Smith BC, Marletta MA. Mechanisms of S-nitrosothiol formation and selectivity in nitric oxide signaling. Current Opinion in Chemical Biology. 2012;**16**:498-506

[62] Ward CA, Giles WR. Ionic mechanism of the effects of hydrogen peroxide in rat ventricular myocytes. The Journal of Physiology. 1997;**500**:631-642

[63] Ma JH, Luo AT, Zhang PH. Effect of hydrogen peroxide on persistent sodium current in Guinea pig ventricular myocytes. Acta Pharmacologica Sinica. 2005;**26**:828-834

[64] Light PE, Wallace CHR, Dyck JRB. Constitutively active adenosine monophosphate-activated protein kinase regulates voltage-gated sodium channels in ventricular myocytes. Circulation. 2003;**107**:1962-1965

[65] Mazzocchi G, Sommese L, Palomeque J, Felice JI, Di Carlo MN, Fainstein D, et al. Phospholamban ablation rescues the enhanced propensity to arrhythmias of mice with CaMKII-constitutive phosphorylation

of RyR2 at site S2814. The Journal of Physiology. 2016;**594**:3005-3030

[66] Bogeski I, Kappl R, Kummerow C, Gulaboski R, Hoth M, Niemeyer BA. Redox regulation of calcium ion channels: Chemical and physiological aspects. Cell Calcium. 2011;**50**:407-423

[67] Bogeski I, Niemeyer BA. Redox regulation of ion channels. Antioxidants & Redox Signaling. 2014;**21**:859-862

[68] Zima AV, Blatter LA. Redox regulation of cardiac calcium channels and transporters. Cardiovascular Research. 2006;**71**:310-321

[69] Sag CM, Wagner S, Maier LS. Role of oxidants on calcium and sodium movement in healthy and diseased cardiac myocytes. Free Radical Biology & Medicine. 2013;**63**:338-349

[70] Yang K-C, Kyle JW, Makielski JC, Dudley SC. Mechanisms of sudden cardiac death. Circulation Research. 2015;**116**:1937-1955

[71] Jeong E-M, Liu M, Sturdy M, Gao G, Varghese ST, Sovari AA, et al. Metabolic stress, reactive oxygen species, and arrhythmia. Journal of Molecular and Cellular Cardiology. 2012;**52**:454-463

[72] Svoboda LK, Reddie KG, Zhang L, Vesely ED, Williams ES, Schumacher SM, et al. Redox-sensitive sulfenic acid modification regulates surface expression of the cardiovascular voltage-gated potassium channel Kv1.5. Circulation Research. 2012;**111**:842-853

[73] Slominski RM, Reiter RJ, Schlabritz-Loutsevitch N, Ostrom RS, Slominski AT. Melatonin membrane receptors in peripheral tissues: Distribution and functions. Molecular and Cellular Endocrinology. 2012;**351**:152-166

[74] Jockers R, Delagrange P, Dubocovich ML, Markus RP, Renault N, Tosini G, et al. Update on melatonin receptors: IUPHAR review 20. British Journal of Pharmacology. 2016;**173**:2702-2725

[75] Stauch B, Johansson LC, McCorvy JD, Patel N, Han GW, Huang XP, et al. Structural basis of ligand recognition at the human MT1 melatonin receptor. Nature. 2019;**569**:284-288

[76] Johansson LC, Stauch B, McCorvy JD, Han GW, Patel N, Huang XP, et al. XFEL structures of the human MT2 melatonin receptor reveal the basis of subtype selectivity. Nature. 2019;**569**:289-292

[77] Benleulmi-Chaachoua A, Chen L, Sokolina K, Wong V, Jurisica I, Emerit MB, et al. Protein interactome mining defines melatonin MT1 receptors as integral component of presynaptic protein complexes of neurons. Journal of Pineal Research. 2016;**60**:95-108

[78] Jockers R, Maurice P, Boutin JA, Delagrange P. Melatonin receptors, heterodimerization, signal transduction and binding sites: What's new? British Journal of Pharmacology. 2008;**154**:1182-1195

[79] Cecon E, Oishi A, Jockers R. Melatonin receptors: Molecular pharmacology and signalling in the context of system bias. British Journal of Pharmacology. 2018;**175**:3263-3280

[80] Oishi A, Karamitri A, Gerbier R, Lahuna O, Ahmad R, Jockers R. Orphan GPR61, GPR62 and GPR135 receptors and the melatonin MT 2 receptor reciprocally modulate their signaling functions. Scientific Reports. 2017;**7**:8990

[81] Hibino H, Inanobe A, Furutani K, Murakami S, Findlay I, Kurachi Y. Inwardly rectifying potassium channels: Their structure, function, and

physiological roles. Physiological Reviews. 2010;**90**:291-366

[82] Baba K, Benleulmi-Chaachoua A, Journé AS, Kamal M, Guillaume JL, Dussaud S, et al. Heteromeric MT1/MT2 melatonin receptors modulate photoreceptor function. Science Signaling. 2013;**6**:ra89-ra89

[83] Prado NJ, Egan Beňová T, Diez ER, Knezl V, Lipták B, Ponce Zumino AZ, et al. Melatonin receptor activation protects against low potassium-induced ventricular fibrillation by preserving action potentials and connexin-43 topology in isolated rat hearts. Journal of Pineal Research. 2019;**67**:e12605

[84] Boutin JA, Ferry G. Is there sufficient evidence that the melatonin binding site MT3 is quinone reductase 2? The Journal of Pharmacology and Experimental Therapeutics. 2019;**368**:59-65

[85] Zhang HM, Zhang Y. Melatonin: A well-documented antioxidant with conditional pro-oxidant actions. Journal of Pineal Research. 2014;**57**:131-146

[86] Boutin JA, Marcheteau E, Hennig P, Moulharat N, Berger S, Delagrange P, et al. MT3/QR2 melatonin binding site does not use melatonin as a substrate or a co-substrate. Journal of Pineal Research. 2008;**45**:524-531

[87] Mailliet F, Ferry G, Vella F, Berger S, Cogé F, Chomarat P, et al. Characterization of the melatoninergic MT3 binding site on the NRH:Quinone oxidoreductase 2 enzyme. Biochemical Pharmacology. 2005;**71**:74-88

[88] Pandi-Perumal SR, Trakht I, Srinivasan V, Spence DW, Maestroni GJMM, Zisapel N, et al. Physiological effects of melatonin: Role of melatonin receptors and signal transduction pathways. Progress in Neurobiology. 2008;**85**:335-353

[89] Gonano LA, Sepúlveda M, Rico Y, Kaetzel M, Valverde CA, Dedman J, et al. Calcium-calmodulin kinase II mediates digitalis-induced arrhythmias. Circulation: Arrhythmia and Electrophysiology. 2011;**4**:947-957

[90] Squecco R, Tani A, Zecchi-Orlandini S, Formigli L, Francini F. Melatonin affects voltage-dependent calcium and potassium currents in MCF-7 cell line cultured either in growth or differentiation medium. European Journal of Pharmacology. 2015;**758**:40-52

[91] Wong R, Steenbergen C, Murphy E. Mitochondrial permeability transition pore and calcium handling. In: Mitochondrial Bioenergetics. Methods in Molecular Biology (Methods and Protocols). Humana Press. 2012;**810**. ISBN 978-1-61779-381-3

[92] Valverde CA, Kornyeyev D, Mattiazzi AR, Escobar AL. Reperfusion after ichemia causes cytosolic calcium overload due to rapid calcium release from the sarcoplasmic reticulum. Biophysical Journal. 2009;**96**:550a

[93] Yeung HM, Hung MW, Fung ML. Melatonin ameliorates calcium homeostasis in myocardial and ischemia-reperfusion injury in chronically hypoxic rats. Journal of Pineal Research. 2008;**45**:373-382

[94] Valverde CA, Mattiazzi A, Escobar AL. Camkii exacerbates calcium waves during reperfusion of ischemic heart. Biophysical Journal. 2014;**106**:322a

[95] Baumeister P, Quinn TA. Altered calcium handling and ventricular arrhythmias in acute ischemia. Clinical Medicine Insights: Cardiology. 2016;**10s1**:CMC.S39706

[96] ter Keurs HEDJ, Boyden PA. Calcium and Arrhythmogenesis. Physiological Reviews. 2007;**87**:457-506

[97] Prado NJ, Casarotto M, Calvo JP, Mazzei L, Ponce Zumino AZ, García IM, et al. Antiarrhythmic effect linked to melatonin cardiorenal protection involves AT 1 reduction and Hsp70-VDR increase. Journal of Pineal Research. 2018;**65**:e12513

[98] Xu W, Cai S-YY, Zhang Y, Wang Y, Ahammed GJ, Xia X-JJ, et al. Melatonin enhances thermotolerance by promoting cellular protein protection in tomato plants. Journal of Pineal Research. 2016;**61**:457-469

[99] Rezzani R, Rodella LF, Bonomini F, Tengattini S, Bianchi R, Reiter RJ. Beneficial effects of melatonin in protecting against cyclosporine A-induced cardiotoxicity are receptor mediated. Journal of Pineal Research. 2006;**41**:288-295

[100] Tas U, Ogeturk M, Kuloglu T, Sapmaz HI, Kocaman N, Zararsiz I, et al. HSP70 immune reactivity and TUNEL positivity in the liver of toluene-inhaled and melatonin-treated rats. Toxicology and Industrial Health. 2013;**29**:514-522

[101] Yoon YM, Kim HJ, Lee JH, Lee SH. Melatonin enhances mitophagy by upregulating expression of heat shock 70 kDa protein 1 L in human mesenchymal stem cells under oxidative stress. International Journal of Molecular Sciences. 2019;**20**:4545

[102] Motawi TK, Ahmed SA, A Hamed M, El-Maraghy SA, M Aziz W. Melatonin and/or rowatinex attenuate streptozotocin-induced diabetic renal injury in rats. Journal of Biomedical Research. 2019;**33**:113-121

[103] Fang N, Hu C, Sun W, Xu Y, Gu Y, Wu L, et al. Identification of a novel melatonin-binding nuclear receptor: Vitamin D receptor. Journal of Pineal Research. 2020;**68**:e12618

[104] Morciano G, Giorgi C, Bonora M, Punzetti S, Pavasini R, Wieckowski MR, et al. Molecular identity of the mitochondrial permeability transition pore and its role in ischemia-reperfusion injury. Journal of Molecular and Cellular Cardiology. 2015;**78**:142-153

[105] Hacışevki A, Baba B. An overview of melatonin as an antioxidant molecule: a biochemical approach. In: Melatonin—Molecular Biology, Clinical and Pharmaceutical Approaches. Rijeka: IntechOpen; 2018

[106] Ganie SA, Dar TA, Bhat AH, Dar KB, Anees S, Zargar MA, et al. Melatonin: A potential anti-oxidant therapeutic agent for mitochondrial dysfunctions and related disorders. Rejuvenation Research. 2016;**19**:21-40

[107] Das N, Mandala A, Naaz S, Giri S, Jain M, Bandyopadhyay D, et al. Melatonin protects against lipid-induced mitochondrial dysfunction in hepatocytes and inhibits stellate cell activation during hepatic fibrosis in mice. Journal of Pineal Research. 2017;**62**:e12404

[108] Donato AJ, Morgan RG, Walker AE, Lesniewski LA. Cellular and molecular biology of aging endothelial cells. Journal of Molecular and Cellular Cardiology. 2015;**89**:122-135

[109] Xu S, Pi H, Zhang L, Zhang N, Li YM, Zhang H, et al. Melatonin prevents abnormal mitochondrial dynamics resulting from the neurotoxicity of cadmium by blocking calcium-dependent translocation of Drp1 to the mitochondria. Journal of Pineal Research. 2016;**60**:291-302

[110] Paradies G, Paradies V, Ruggiero FM, Petrosillo G. Mitochondrial bioenergetics decay in aging: Beneficial effect of melatonin. Cellular and Molecular Life Sciences. 2017;**74**:3897-3911

[111] Ma Z, Xin Z, Di W, Yan X, Li X, Reiter RJ, et al. Melatonin and

mitochondrial function during ischemia/reperfusion injury. Cellular and Molecular Life Sciences. 2017;**74**:3989-3998

[112] Petrosillo G, Di Venosa N, Pistolese M, Casanova G, Tiravanti E, Colantuono G, et al. Protective effect of melatonin against mitochondrial dysfunction associated with cardiac ischemia- reperfusion: Role of cardiolipin. The FASEB Journal. 2006;**20**:269-276

[113] Petrosillo G, Moro N, Paradies V, Ruggiero FM, Paradies G. Increased susceptibility to Ca^{2+}-induced permeability transition and to cytochrome c release in rat heart mitochondria with aging: Effect of melatonin. Journal of Pineal Research. 2010;**45**:340-346

[114] Kang JW, Hong JM, Lee SM. Melatonin enhances mitophagy and mitochondrial biogenesis in rats with carbon tetrachloride-induced liver fibrosis. Journal of Pineal Research. 2016;**60**:383-393

[115] Pei HF, Hou JN, Wei FP, Xue Q, Zhang F, Peng CF, et al. Melatonin attenuates postmyocardial infarction injury via increasing Tom70 expression. Journal of Pineal Research. 2017;**62**:e12371

[116] Fan ACY, Young JC. Function of cytosolic chaperones in Tom70-mediated mitochondrial import. Protein and Peptide Letters. 2011;**18**:122-131

[117] Lochner A, Marais E, Huisamen B. Melatonin and cardioprotection against ischaemia/reperfusion injury: What's new? A review. Journal of Pineal Research. 2018;**65**:e12490

[118] Matsushima S, Sadoshima J. The role of sirtuins in cardiac disease. The American Journal of Physiology-Heart and Circulatory Physiology. 2015;**309**:H1375-H1389

[119] D'Onofrio N, Servillo L, Balestrieri ML. SIRT1 and SIRT6 signaling pathways in cardiovascular disease protection. Antioxidants & Redox Signaling. 2018;**28**:711-732

[120] Yu L, Sun Y, Cheng L, Jin Z, Yang Y, Zhai M, et al. Melatonin receptor-mediated protection against myocardial ischemia/reperfusion injury: Role of SIRT1. Journal of Pineal Research. 2014;**57**:228-238

[121] Yu L, Liang H, Dong X, Zhao G, Jin Z, Zhai M, et al. Reduced silent information regulator 1 signaling exacerbates myocardial ischemia-reperfusion injury in type 2 diabetic rats and the protective effect of melatonin. Journal of Pineal Research. 2015;**59**:376-390

[122] Mayo JC, Sainz RM, González Menéndez P, Cepas V, Tan DX, Reiter RJ. Melatonin and sirtuins: A "not-so unexpected" relationship. Journal of Pineal Research. 2017;**62**:e12391

[123] Yu L, Gong B, Duan W, Fan C, Zhang J, Li Z, et al. Melatonin ameliorates myocardial ischemia/reperfusion injury in type 1 diabetic rats by preserving mitochondrial function: Role of AMPK-PGC-1α-SIRT3 signaling. Scientific Reports. 2017;**7**:41337

[124] Chua S, Lee F-YY, Chiang H-JJ, Chen K-HH, Lu H-II, Chen Y-TTY-LL, et al. The cardioprotective effect of melatonin and exendin-4 treatment in a rat model of cardiorenal syndrome. Journal of Pineal Research. 2016;**61**:438-456

[125] Majidinia M, Sadeghpour A, Mehrzadi S, Reiter RJ, Khatami N, Yousefi B. Melatonin: A pleiotropic molecule that modulates DNA damage response and repair pathways. Journal of Pineal Research. 2017;**63**:e12416

[126] Ben SI, Mies F, Naeije R, Shlyonsky V. Melatonin

down-regulates volume-sensitive chloride channels in fibroblasts. Pflügers Archiv: European Journal of Physiology. 2012;**464**:273-285

[127] Diez ER, Prados LV, Carrión A, Ponce ZAZ, Miatello RM. A novel electrophysiologic effect of melatonin on ischemia/reperfusion-induced arrhythmias in isolated rat hearts. Journal of Pineal Research. 2009;**46**:155-160

[128] Diez ER, Prado NJ, Carrión AM, Petrich ER, Ponce Zumino AZ, Miatello RM. Electrophysiological effects of tamoxifen: Mechanism of protection against reperfusion arrhythmias in isolated rat hearts. Journal of Cardiovascular Pharmacology. 2013;**62**:184-191

[129] Varga Z, Panyi G, Péter M, Pieri C, Csécsei G, Damjanovich S, et al. Multiple binding sites for melatonin on Kv1.3. Biophysical Journal. 2001;**80**:1280-1297

[130] Hablitz LM, Molzof HE, Abrahamsson KE, Cooper JM, Prosser RA, Gamble KL. GIRK channels mediate the nonphotic effects of exogenous melatonin. The Journal of Neuroscience. 2015;**35**:14957-14965

[131] Nelson CS, Marino JL, Allen CN. Melatonin receptors activate heteromeric G-protein coupled Kir3 channels. Neuroreport. 1996;**7**:717-720

[132] Huang Y, Li Y, Leng Z. Melatonin inhibits GABAergic neurons in the hypothalamus consistent with a reduction in wakefulness. Neuroreport. 2020;**31**:92-98

[133] Waseem M, Tabassum H, Parvez S. Melatonin modulates permeability transition pore and 5-hydroxydecanoate induced KATP channel inhibition in isolated brain mitochondria. Mitochondrion. 2016;**31**:1-8

[134] Mohammadi F, Shakiba S, Mehrzadi S, Afshari K, Rahimnia AH, Dehpour AR. Anticonvulsant effect of melatonin through ATP-sensitive channels in mice. Fundamental & Clinical Pharmacology. 2020;**34**:148-155

[135] Brown DA, Aon MA, Frasier CR, Sloan RC, Maloney AH, Anderson EJ, et al. Cardiac arrhythmias induced by glutathione oxidation can be inhibited by preventing mitochondrial depolarization. Journal of Molecular and Cellular Cardiology. 2010;**48**:673-679

[136] Nakaya H. Role of ATP-sensitive K^+ channels in cardiac arrhythmias. Journal of Cardiovascular Pharmacology and Therapeutics. 2014;**19**:237-243

[137] Ayar A, Martin DJ, Ozcan M, Kelestimur H. Melatonin inhibits high voltage activated calcium currents in cultured rat dorsal root ganglion neurones. Neuroscience Letters. 2001;**313**:73-77

[138] Choi TY, Kwon JE, Durrance ES, Jo SH, Choi SY, Kim KT. Melatonin inhibits voltage-sensitive Ca^{2+} channel-mediated neurotransmitter release. Brain Research. 2014;**1557**:34-42

[139] Escames G, Macías M, León J, García J, Khaldy H, Martín M, et al. Calcium-dependent effects of melatonin inhibition of glutamatergic response in rat striatum. Journal of Neuroendocrinology. 2001;**13**:459-466

[140] Mei YA, Lee PPN, Wei H, Zhang ZH, Pang SF. Melatonin and its analogs potentiate the nifedipine-sensitive high-voltage-activated calcium current in the chick embryonic heart cells. Journal of Pineal Research. 2001;**30**:13-21

[141] Mück AO, Seeger H, Bartsch C, Lippert TH. Does melatonin affect calcium influx in human aortic smooth muscle cells and estradiol-mediated

calcium antagonism? Journal of Pineal Research. 1996;**20**:145-147

[142] Chen LD, Kumar P, Reiter RJ, Tan DX, Chamber JP, Manchester LC, et al. Melatonin reduces 3H-nitrendipine binding in the heart. Proceedings of the Society for Experimental Biology and Medicine. 1994;**207**:34-37

[143] Xu Z, Wu Y, Zhang Y, Zhang H, Shi L. Melatonin activates BKCa channels in cerebral artery myocytes via both direct and MT receptor/PKC-mediated pathway. European Journal of Pharmacology. 2019;**842**:177-188

[144] Al Dera H, Alassiri M, Eleawa SM, AlKhateeb MA, Hussein AM, Dallak M, et al. Melatonin improves memory deficits in rats with cerebral hypoperfusion, possibly, through decreasing the expression of small-conductance Ca^{2+}-activated K^{+} channels. Neurochemical Research. 2019;**44**:1851-1868

[145] Ovali MA, Uzun M. The effects of melatonin administration on KCNQ and KCNH2 gene expressions and QTc interval in pinealectomised rats. Cellular and Molecular Biology. 2017;**63**:45-50

[146] Arntz HR, Willich SN, Oeff M, Brüggemann T, Stern R, Heinzmann A, et al. Circadian variation of sudden cardiac death reflects age-related variability in ventricular fibrillation. Circulation. 1993;**88**:2284-2289

[147] Tribulova N, Szeiffova Bacova B, Benova T, Viczenczova C. Can we protect from malignant arrhythmias by modulation of cardiac cell-to-cell coupling? Journal of Electrocardiology. 2015;**48**:434-440

[148] Babiker F, Al-Jarallah A, Al-Awadi M. Effects of cardiac hypertrophy, diabetes, aging, and pregnancy on the Cardioprotective effects of Postconditioning in male and female rats. Cardiology Research and Practice. 2019;**2019**:3403959

[149] Carro J, Pueyo E, Rodríguez Matas JF. A response surface optimization approach to adjust ionic current conductances of cardiac electrophysiological models. Application to the study of potassium level changes. PLoS One. 2018;**13**:e0204411

[150] Johnson R, Camelliti P. Role of non-Myocyte gap junctions and Connexin Hemichannels in cardiovascular health and disease: Novel therapeutic targets? International Journal of Molecular Sciences. 2018;**19**:866

[151] Pk W, Whelton PK, Carey RM, Aronow WS, Ovbiagele B, Casey DE, et al. Guideline for the prevention, detection, evaluation, and management of high blood pressure in adults. Journal of American College of Cardiology. 2017;**283**

[152] Steinberg C, Laksman ZWM, Krahn AD. Sudden cardiac death: A reappraisal. Trends in Cardiovascular Medicine. 2016;**26**:709-719

[153] Ibanez B, James S, Agewall S, Antunes MJ, Bucciarelli-Ducci C, Bueno H, et al. 2017 ESC guidelines for the management of acute myocardial infarction in patients presenting with ST-segment elevation. European Heart Journal. 2018;**39**:119-177

[154] Roth GA, Huffman MD, Moran AE, Feigin V, Mensah GA, Naghavi M, et al. Global and regional patterns in cardiovascular mortality from 1990 to 2013. Circulation. 2015;**132**:1667-1678

[155] Ambale-Venkatesh B, Yang X, Wu CO, Liu K, Hundley WG, McClelland R, et al. Cardiovascular event prediction by machine learning. Circulation Research. 2017;**121**:1092-1101

[156] Adabag AS, Luepker RV, Roger VL, Gersh BJ. Sudden cardiac death: Epidemiology and risk factors. Nature Reviews. Cardiology. 2010;7:216-225

[157] Li Y, Nantsupawat T, Tholakanahalli V, Adabag S, Wang Z, Benditt DG, et al. Characteristics and periodicity of sustained ventricular tachyarrhythmia events in a population of military veterans with implantable cardioverter defibrillator. Journal of Interventional Cardiac Electrophysiology. 2019. DOI: 10.1007/s10840-019-00569-0. [Epub ahead of print]

[158] Tang PT, Shenasa M, Boyle NG. Ventricular arrhythmias and sudden cardiac death. Cardiac Electrophysiology Clinics. 2017;**9**:693-708

[159] Tan D-X, Manchester LC, Reiter RJ, Qi W, Kim SJ, El-Sokkary GH. Ischemia/reperfusion-induced arrhythmias in the isolated rat heart: Prevention by melatonin. Journal of Pineal Research. 1998;**25**:184-191

[160] Lagneux C, Joyeux M, Demenge P, Ribuot C, Godin-Ribuot D. Protective effects of melatonin against ischemia-reperfusion injury in the isolated rat heart. Life Sciences. 2000;**66**:503-509

[161] Szárszoi O, Asemu G, Vaněček J, Ošt'ádal B, Kolář F. Effects of melatonin on ischemia and reperfusion injury of the rat heart. Cardiovascular Drugs and Therapy. 2001;**15**:251-257

[162] Kaneko S, Okumura K, Numaguchi Y, Matsui H, Murase K, Mokuno S, et al. Melatonin scavenges hydroxyl radical and protects isolated rat hearts from ischemic reperfusion injury. Life Sciences. 2000;**67**:101-112

[163] Lee YM, Chen HR, Hsiao G, Sheu JR, Wang JJ, Yen MH. Protective effects of melatonin on myocardial ischemia/reperfusion injury in vivo. Journal of Pineal Research. 2002;**33**:72-80

[164] Bertuglia S, Reiter RJ. Melatonin reduces ventricular arrhythmias and preserves capillary perfusion during ischemia-reperfusion events in cardiomyopathic hamsters. Journal of Pineal Research. 2007;**42**:55-63

[165] Važan R, Pancza D, Béder I, Styk J. Ischemia-reperfusion injury—Antiarrhythmic effect of melatonin associated with reduced recovering of contractility. General Physiology and Biophysics. 2005;**24**:355-359

[166] Vazan R, Ravingerova T. Protective effect of melatonin against myocardial injury induced by epinephrine. Journal of Physiology and Biochemistry. 2015;**71**:43-49

[167] Dobsak P, Siegelova J, Eicher JC, Jancik J, Svacinova H, Vasku J, et al. Melatonin protects against ischemia-reperfusion injury and inhibits apoptosis in isolated working rat heart. Pathophysiology. 2003;**9**:179-187

[168] Diez ER, Renna NF, Prado NJ, Lembo C, Ponce Zumino AZ, Vazquez-Prieto M, et al. Melatonin, given at the time of reperfusion, prevents ventricular arrhythmias in isolated hearts from fructose-fed rats and spontaneously hypertensive rats. Journal of Pineal Research. 2013;**55**:166-173

[169] Benova T, Viczenczova C, Radosinska J, Bacova B, Knezl V, Dosenko V, et al. Melatonin attenuates hypertension-related proarrhythmic myocardial maladaptation of connexin-43 and propensity of the heart to lethal arrhythmias. Canadian Journal of Physiology and Pharmacology. 2013;**91**:633-639

[170] Lee S, Kang E, Yoo KD, Choi Y, Kim DK, Joo KW, et al. Lower serum potassium associated with increased

mortality in dialysis patients: A nationwide prospective observational cohort study in Korea. PLoS One. 2017;**12**:e0171842

[171] Pun PH, Goldstein BA, Gallis JA, Middleton JP, Svetkey LP. Serum potassium levels and risk of sudden cardiac death among patients with chronic kidney disease and significant coronary artery disease. Kidney International Reports. 2017;**2**:1122-1131

[172] Osadchii OE. Electrophysiological determinants of arrhythmic susceptibility upon endocardial and epicardial pacing in Guinea-pig heart. Acta Physiologica. 2012;**205**:494-506

[173] Osadchii OE. Impact of hypokalemia on electromechanical window, excitation wavelength and repolarization gradients in guinea-pig and rabbit hearts. PLoS One. 2014;**9**:e105599

[174] Osadchii OE. Effects of antiarrhythmics and hypokalemia on the rate adaptation of cardiac repolarization. Scandinavian Cardiovascular Journal. 2018;**52**:218-226

[175] Xie Y, Sato D, Garfinkel A, Qu Z, Weiss JN. So little source, so much sink: Requirements for after depolarizations to propagate in tissue. Biophysical Journal. 2010;**99**:1408-1415

[176] Osadchii OE. Mechanisms of hypokalemia-induced ventricular arrhythmogenicity. Fundamental & Clinical Pharmacology. 2010;**24**:547-559

[177] Ek-Vitorín J, Pontifex T, Burt J, Ek-Vitorín JF, Pontifex TK, Burt JM. Cx43 channel gating and permeation: Multiple phosphorylation-dependent roles of the carboxyl terminus. International Journal of Molecular Sciences. 2018;**19**:1659

[178] Palatinus JA, O'Quinn MP, Barker RJ, Harris BS, Jourdan J, Gourdie RG. ZO-1 determines adherens and gap junction localization at intercalated disks. The American Journal of Physiology: Heart and Circulatory Physiology. 2011;**300**:H583-H594

[179] Iyyathurai J, Himpens B, Bultynck G, D'hondt C. Calcium wave propagation triggered by local mechanical stimulation as a method for studying gap junctions and hemichannels. Methods in Molecular Biology (Clifton, NJ). 2016;**1437**:203-211

[180] Dhein S, Seidel T, Salameh A, Jozwiak J, Hagen A, Kostelka M, et al. Remodeling of cardiac passive electrical properties and susceptibility to ventricular and atrial arrhythmias. Frontiers in Physiology. 2014;**5**:424

[181] Xie Y, Garfinkel A, Camelliti P, Kohl P, Weiss JN, Qu Z. Effects of fibroblast-myocyte coupling on cardiac conduction and vulnerability to reentry: A computational study. Heart Rhythm. 2009;**6**:1641-1649

[182] Remo BF, Qu J, Volpicelli FM, Giovannone S, Shin D, Lader J, et al. Phosphatase-resistant gap junctions inhibit pathological remodeling and prevent arrhythmias. Circulation Research. 2011;**108**:1459-1466

[183] Hoppe LK, Muhlack DC, Koenig W, Carr PR, Brenner H, Schöttker B. Association of abnormal serum potassium levels with arrhythmias and cardiovascular mortality: A systematic review and meta-analysis of observational studies. Cardiovascular Drugs and Therapy. 2018;**32**:197-212

[184] Trenor B, Cardona K, Romero L, Gomez JF, Saiz J, Rajamani S, et al. Pro-arrhythmic effects of low plasma [K^+] in human ventricle: An illustrated review. Trends in Cardiovascular Medicine. 2018;**28**:233-242

[185] Ishigaki S, Ohashi N, Isobe S, Tsuji N, Iwakura T, Ono M, et al. Impaired endogenous nighttime melatonin secretion relates to intrarenal renin-angiotensin system activation and renal damage in patients with chronic kidney disease. Clinical and Experimental Nephrology. 2016;**20**:878-884

[186] Zetner D, Andersen L, Rosenberg J. Pharmacokinetics of alternative administration routes of melatonin: A systematic review. Drug Research (Stuttgart). 2015;**66**:169-173

[187] Andersen LPH, Gögenur I, Rosenberg J, Reiter RJ. The safety of melatonin in humans. Clinical Drug Investigation. 2016;**36**:169-175

[188] Gul-Kahraman K, Yilmaz-Bozoglan M, Sahna E. Physiological and pharmacological effects of melatonin on remote ischemic perconditioning after myocardial ischemia-reperfusion injury in rats: Role of Cybb, Fas, NfκB, Irisin signaling pathway. Journal of Pineal Research. 2019;**67**:e12589

[189] Sahna E, Olmez E, Acet A. Effects of physiological and pharmacological concentrations of melatonin on ischemia-reperfusion arrhythmias in rats: Can the incidence of sudden cardiac death be reduced? Journal of Pineal Research. 2002;**32**:194-198

[190] Dominguez-Rodriguez A, Abreu-Gonzalez P, Reiter RJ. Circadian variation in acute myocardial infarction size: Likely involvement of the melatonin and suprachiasmatic nuclei. International Journal of Cardiology. 2017;**235**:191

[191] Nagibin V, Egan Benova T, Viczenczova C, Szeiffova Bacova B, Dovinova I, Barancik M, et al. Ageing related down-regulation of myocardial connexin-43 and up-regulation of MMP-2 may predict propensity to atrial fibrillation in experimental animals. Physiological Research. 2016;**65**(Suppl 1):S91-S100

[192] Sedova KA, Bernikova OG, Cuprova JI, Ivanova AD, Kutaeva GA, Pliss MG, et al. Association between antiarrhythmic, electrophysiological, and Antioxidative effects of melatonin in ischemia/reperfusion. International Journal of Molecular Sciences. 2019;**20**:6331

[193] Al-Khatib SM, Stevenson WG, Ackerman MJ, Bryant WJ, Callans DJ, Curtis AB, et al. AHA/ACC/HRS guideline for management of patients with ventricular arrhythmias and the prevention of sudden cardiac death: Executive summary: A report of the American College of Cardiology/American Heart Association task force on clinical practice Gui. Heart Rhythm. 2017;**15**:e190-e252

[194] Li J, Hu D, Song X, Han T, Gao Y, Xing Y. The role of biologically active ingredients from natural drug treatments for arrhythmias in different mechanisms. BioMed Research International. 2017;**2017**:1-10

[195] Al-Gobari M, Al-Aqeel S, Gueyffier F, Burnand B. Effectiveness of drug interventions to prevent sudden cardiac death in patients with heart failure and reduced ejection fraction: An overview of systematic reviews. BMJ Open. 2018;**8**:e021108

[196] Erath JW, Hohnloser SH. Drugs to prevent sudden cardiac death. International Journal of Cardiology. 2017;**237**:22-24

[197] Zanoboni A, Zanoboni-Muciaccia W. Experimental hypertension in pinealectomized rats. Life Sciences. 1967;**6**:2327-2331

[198] Holmes SW, Sugden D. Proceedings: The effect of melatonin on pinealectomy-induced hypertension in the rat. British Journal of Pharmacology. 1976;**56**:360P-361P

[199] De Farias TDSM, De Oliveira AC, Andreotti S, Do Amaral FG, Chimin P, De Proença ARA, et al. Pinealectomy interferes with the circadian clock genes expression in white adipose tissue. Journal of Pineal Research. 2015;**58**:251-261

[200] Simko F, Pechanova O. Remodelling of the heart and vessels in experimental hypertension: Advances in protection. Journal of Hypertension. 2010;**28**(Suppl 1):S1-S6

[201] Russcher M, Koch B, Nagtegaal E, van der Putten K, ter Wee P, Gaillard C. The role of melatonin treatment in chronic kidney disease. Frontiers in Bioscience (Landmark Ed). 2012;**17**: 2644-2656

[202] Campos LA, Cipolla-Neto J, Amaral FG, Michelini LC, Bader M, Baltatu OC. The angiotensin-melatonin Axis. International Journal of Hypertension. 2013;**2013**:1-7

[203] Ishigaki S, Ohashi N, Matsuyama T, Isobe S, Tsuji N, Iwakura T, et al. Melatonin ameliorates intrarenal renin–angiotensin system in a 5/6 nephrectomy rat model. Clinical and Experimental Nephrology. 2018;**22**:539-549

[204] Franczyk-Skóra B, Gluba-Brzózka A, Wranicz JK, Banach M, Olszewski R, Rysz J. Sudden cardiac death in CKD patients. International Urology and Nephrology. 2015;**47**: 971-982

[205] Pun PH. The interplay between CKD, sudden cardiac death, and ventricular arrhythmias. Advances in Chronic Kidney Disease. 2014;**21**:480-488

[206] Bonato FOB, Canziani MEF, Bonato FOB, Canziani MEF. Ventricular arrhythmia in chronic kidney disease patients. Jornal Brasileiro de Nefrologia. 2017;**39**:186-195

[207] Bonato FOB, Watanabe R, Lemos MM, Cassiolato JL, Wolf M, Canziani MEF. Asymptomatic ventricular arrhythmia and clinical outcomes in chronic kidney disease: A pilot study. The Journal CardioRenal Medicine. 2016;**7**:66-73

[208] Reddan DN, Szczech L, Bhapkar MV, Moliterno DJ, Califf RM, Ohman EM, et al. Renal function, concomitant medication use and outcomes following acute coronary syndromes. Nephrology, Dialysis, Transplantation. 2005;**20**:2105-2112

[209] Diez ER, Altamirano LB, García IM, Mazzei L, Prado NJ, Fornes MW, et al. Heart remodeling and ischemia–reperfusion arrhythmias linked to myocardial vitamin D receptors deficiency in obstructive nephropathy are reversed by Paricalcitol. Journal of Cardiovascular Pharmacology and Therapeutics. 2015;**20**:211-220

[210] Mazzei L, Docherty NG, Manucha W. Mediators and mechanisms of heat shock protein 70 based cytoprotection in obstructive nephropathy. Cell Stress & Chaperones. 2015;**20**:893-906

[211] Lutz W, Kohno K, Kumar R. The role of heat shock protein 70 in vitamin D receptor function. Biochemical and Biophysical Research Communications. 2001;**282**:1211-1219

[212] Ferder M, Inserra F, Manucha W, Ferder L. The world pandemic of vitamin D deficiency could possibly be explained by cellular inflammatory response activity induced by the renin-angiotensin system. The American Journal of Physiology-Cell Physiology. 2013;**304**:C1027-C1039

[213] García IM, Altamirano L, Mazzei L, Fornés M, Cuello-Carrión FD, Ferder L, et al. Vitamin D receptor-modulated Hsp70/AT1 expression may

protect the kidneys of SHRs at the structural and functional levels. Cell Stress & Chaperones. 2014;**19**:479-491

[214] McElwee SK, Velasco A, Doppalapudi H. Mechanisms of sudden cardiac death. Journal of Nuclear Cardiology. 2016;**23**:1368-1379

[215] van den Top M, Buijs RM, Ruijter JM, Delagrange P, Spanswick D, Hermes ML. Melatonin generates an outward potassium current in rat suprachiasmatic nucleus neurones in vitro independent of their circadian rhythm. Neuroscience. 2001;**107**:99-108

[216] Hou S-W, Zheng P, Sun F-Y. Melatonin inhibits outward delayed rectifier potassium currents in hippocampal CA1 pyramidal neuron via intracellular indole-related domains. Journal of Pineal Research. 2004;**36**:242-249

[217] Yang XF, Miao Y, Ping Y, Wu HJ, Yang XL, Wang Z. Melatonin inhibits tetraethylammonium-sensitive potassium channels of rod ON type bipolar cells via MT2 receptors in rat retina. Neuroscience. 2011;**173**:19-29

Section 4

Controlled Release Melatonin Formulations

Chapter 5

Per Os Administered Modified-Release Solid Formulations of Melatonin: A Review of the Latest Developments Including the Design of Experiments (DoE) Approach

Angeliki Siamidi and Yannis Dotsikas

Abstract

The pineal hormone melatonin (MLT) is a derivative of the amino acid *L*-tryptophan and controls the circadian diurnal rhythm and the seasonal biorhythm. Exogenous administration is aimed at alleviating sleep-related dysfunctions and jet lag, as it decreases sleep-onset latency, increases total sleep time and improves overall sleep quality. Besides these indications, MLT has been shown to have other actions, such as antioxidant, immune enhancement and anticancer. It has also been shown to be useful against cardiovascular, neurological and psychiatric diseases. In the context of this work, a review of the related literature on the modified release of MLT from its *per os* administered formulations is presented, including the utilization of the design of experiments (DoE) for the selection of the optimal composition of melatonin formulations. The chapter offers an account of the recent advantages on MLT's solid dosage forms suitable for treating sleep disorders, referring either to its onset or maintenance.

Keywords: melatonin, circadian rhythm, sleep disorders, modified release, oral solid dosage forms, *per os* administration, experimental design

1. Introduction

Melatonin (MLT) is an indole amide hormone produced by the pineal gland, especially at night time and is mainly involved in the regulation of circadian and circannual rhythms. For clinical purposes (including Alzheimer's disease, insomnia, stroke, depression, Parkinson's disease, migraine, headache, etc.), exogenous MLT could be administered for alleviation of the symptoms. Irrespective of the pathological case, MLT's exogenous administration should mimic the typical nocturnal endogenous MLT levels. The release profile of these drug delivery systems should be in a controlled manner, due to MLT's short half-life of elimination and low bioavailability.

The development of novel pharmaceutical formulations with the optimal release profile is of great importance for cases like melatonin. To this purpose, a series of experiments should be performed, utilizing various excipients at different ratios. These trials could require plenty of working hours, with no guarantee that, indeed, the optimal formulation composition will be reached. Therefore, the employment of a statistical/chemometric approach, such as design of experiments (DoE), can be beneficial for complicated and demanding tasks like the development of modified-release formulations with a minimum number of experiments.

2. Advances on the melatonin *per os* administered modified-release solid formulations

Scientific research has been conducted for the purpose of MLT evaluation not only towards medicine and clinical evaluation but also towards pharmaceutics [1–3]. Many researchers have studied and managed to produce immediate-release formulations of MLT (tablets, creams, sublingual sprays, nasal preparations, injectables, etc.), in order to facilitate sleep-onset problems. MLT modified-release formulations are clinically more useful in initiating and maintaining sleep in elderly insomniacs than those designed for immediate release. More in particular, the sustained-release dosage form which delivers MLT in a time period over 8 h is of clinical value for those who have disordered circadian rhythms [4, 5]. Therefore, the modified release of MLT from oral solid dosage forms that alter the time and/or the rate of MLT release may provide an alternative for MLT delivery and be useful in the treatment of circadian rhythmic disorders, like insomnia, jet lag, seasonal affective disease, shift work syndrome, etc. [6]. The recent advantages of MLT modified-release *per oral* solid formulations are reviewed below and summarized in **Table 1**.

Aiming at this mode of release of MLT, a series of hydrophilic matrix tablets has been prepared and tested in vitro. The tablets comprised of combinations of excipients (hydroxypropyl methylcellulose K 15 M (HPMC), low-viscosity sodium alginate, Avicel PH 102, etc.) and a variety of cyclodextrins (MLT (guest)-cyclodextrin (host) complexes in 1:1 ratio). The release studies that were performed in two dissolution media (acidic pH 1.2 and basic pH 7.4) suggested that melatonin was released faster from the MLT-cyclodextrin complexes than from the matrix systems possibly due to their increased solubility [7]. In another investigation, a rather unexploited biomaterial for applications in the design of drug delivery systems, the algal sulphated polysaccharide ulvan was used as an excipient in MLT solid dosage forms. The dissolution tests showed that the MLT release from the ulvan-based tablets followed a sigmoidal pattern, which denoted that the drug release is controlled by polymer relaxation and/or erosion [8]. In a similar study, hydrophilic matrix tablets with various excipients (hydroxypropyl methylcellulose K15 M, low-viscosity sodium alginate, lactose monohydrate and polyvinylpyrrolidone M.W.: 10.000 and 55.000) were developed and tested in vitro at two dissolution media (pH 1.2 and 7.4) in order to examine the modified-release characteristics of MLT. The objective was to produce a formulation with a quick initial pace, aiming at a satisfactory sleep onset, followed by a prolonged release that could target poor sleep quality problems. The dissolution results indicated that the combination of the excipients with different physicochemical properties could alter the release of MLT from solid matrix systems [9].

Moreover, researchers have developed matrix tablets comprised of common hydrogels (hydroxypropyl methylcellulose and dextran) to study the influence on the release profile of MLT in vitro and liposomes (of 1,2-dipalmitoyl-sn-glycero-3-phosphocholine and dipalmitoylphosphatidylglycerol) incorporating the hormone in order to compare

Drug release behaviour	Delivery system	Excipient (s)	Reference
Modified	Matrix tablets	HPMC K15 M, MCC, sodium alginate and various cyclodextrins	[7]
		Ulvan, HPMC K15 M, low-viscosity sodium alginate, LM, PVP	[8]
		PVP (M.W.: 10.000 and 55.000), sodium alginate (low viscosity 2%), HPMC K15 M, dextran, MCC, LM	[9]
		Dextran, MCC, HPMC K15 M, MCC, LM	[10],
	Liposomes	DPPC, DPPG	[10]
	Nanofibrous electrospun mats incorporated into monolayered and three-layered tablets	PVP (M.W.: 1.300.000), CA (M.W.: 50.000), HPMC K15 M, LM	[11]
	Electrospun nanofibres in capsules	PVP (M.W.: 1.300.000), CA (M.W.: 50.000), hypromellose 2910, PEO (M.W. 900.000 and 400.000)	[12]
	Calcium alginate beads in hard gelatin capsules		[13]
Slow	Matrix tablets	HPMC, Carbopol 971P, MCC, maize starch	[14]
Sustained	Matrix tablets	HPMC, Avicel, Primojel, Ac-Di-Sol, Polyplasdone, Mg stearate, Talc, Cab-O-Sil	[15]
	Solid lipid nanoparticles in hard gelatin capsules	Stearic acid, Epikuron 200, lactose	[16]
Controlled	Matrix tablets	HPMCK15 M, low-viscosity sodium alginate, LM, PVP (M.W.: 10.000 or 55.000)	[17]
	Beads	Sodium alginate, Eudragit® RS100, aluminum tristearate, polyethylene, glycol 400, liquid paraffin	[18]
Delayed	Compression-coated tablets	Dextran, PVP(M.W.: 10.000), ethyl cellulose (45cps), Avicel PH 102, LM, sodium alginate	[19]
Biphasic	Matrix tablets	Dextran, PVP (10.000), ethyl cellulose (45cps), MCC, LM, sodium alginate	[19]
Immediate and sustained	Bilayer tablets	Dextran, ethyl cellulose (45cps), MCC, LM, sodium alginate	[19]
		MT-b-CD, HPMC, Carbopol 971P, MCC, maize starch	[20]
Immediate and controlled	Coated beads in hard gelatin capsule	Core sugar spheres, Aquacoat®, dibutyl sebacate, triethyl citrate, PVP, HPMC	[20]

HPMC: Hydroxypropyl methylcellulose, MCC: Microcrystalline cellulose (avicel PH 102), PVP: Polyvinylpyrrolidone, PEO: Polyethylene oxide, CA: Cellulose acetate, LM: Lactose monohydrate, DPPC: 1,2-dipalmitoyl-sn-glycero-3-phosphocholine, DPPG: Dipalmitoyl-phosphatidylglycerol.

Table 1.
An overview of the recent advantages of MLT modified-release per oral *solid formulations.*

their release profiles. The results indicated that both formulations (liposomal and solid matrix tablets) could be suitable alternatives for treating sleep-onset/maintenance problems [10].

Once more, aiming at the modified release of MLT, another group of researchers studied the MLT release from monolayered and three-layered tablets, incorporating nanofibrous mats composed of cellulose acetate and polyvinylpyrrolidone. The in vitro dissolution release studies of the MLT formulations in simulated gastrointestinal fluids revealed tableting pressure and pH dependence. Comparing the MLT release from the physical mixture tablets and from the nanofibre-based tablets, it was concluded that the release profile was generally slower than the latter, rendering the formulation suitable for both sleep-onset and maintenance dysfunctions [11]. The same group of researchers produced electrospun-MLT loaded nanofibres (with cellulose acetate, polyvinylpyrrolidone and hydroxypropyl methylcellulose, as excipients) and used them to fill hard gelatin and delayed-release (DRcaps™) capsules. The in vitro dissolution results revealed a modified-release profile of MLT from the fabricated matrices in gastrointestinal-like fluids and suggested that the MLT-loaded nanofibrous mats could exhibit a promising profile for treating sleep problems [12].

Calcium alginate beads were also prepared to investigate the MLT modified release. Excipients utilized in their preparation included calcium alginate, polyvinylpyrrolidone (M.W.: 10.000 and 55.000), hydroxypropyl methylcellulose (M.W.: 15.000 and 100.000), lactose monohydrate and, as a surfactant, sodium lauryl sulphate. The in vitro release of melatonin was investigated at two different pHs (acidic pH 1.2 and basic pH 6.8), and the results concluded that the hormone's release from the beads was reversibly proportional to the extent of their expansion, which depends on the molecular weight/viscosity of the biopolymers present in the beads; the higher the molecular weight/viscosity of the hydrogels, the greater the beads swelling and the less the MLT's release [13].

Another group of researchers prepared a slow-release tablet of MLT with varying quantities of hydroxypropyl methylcellulose K15 M and Carbopol 971P, as well as other excipients (microcrystalline cellulose, maize starch, magnesium stearate and purified talc). The formulations developed showed a slow release of MLT during an 8 h period [14].

To the same end, matrix tablets were formulated using hydroxypropyl methylcellulose and tested in vitro in relation to drug release, as a function of polymer viscosity, drug loading, type and amount of disintegrant, lubricant and glidant and aqueous polymeric coating level, and further compared with two commercial products. The release studies showed that as the polymer viscosity increased, the release decreased, and as the coating level increased, an increased lag time was observed [15]. Other researchers have examined the in vivo sustained release of MLT that was incorporated in solid lipid nanoparticles. The results indicated that solid lipid nanoparticles may act as a reservoir, permitting a constant and prolonged MLT release, after oral administration, which may indicate new possibilities for sustained delivery systems [16].

In another research project, controlled-release matrix tablets of MLT were developed by the use of a computer programme, D-optimal experimental design, aiming at affecting its modified release at simulated gastrointestinal media. The careful selection of the excipients (polyvinylpyrrolidone (M.W.: 10.000 and 55.000), hydroxypropyl methylcellulose K15 M and lactose monohydrate) at their appropriate quantity resulted to the optimal solution and the controlled release of melatonin with the minimal number of experiments [17]. Moreover, in another research, polymer-reinforced and polymer-coated alginate beads with various concentrations of polymer (Eudragit® RSI00) and plasticizer (aluminum tristearate) were produced and evaluated in vitro in relation to their controlled-release characteristics as an alternative for oral delivery of MLT. The results indicated that the polymeric reinforcement offered an initial burst release in intestinal fluids, while the coating led to release retardation in both gastric and intestinal fluids. The results also showed that as the polymer concentration increased, the MLT release decreased in the intestinal fluids, due to coated alginate beads disintegration [18].

Researchers have also probed the MLT release from matrix and compression-coated tablets that were comprised of combinations of ethyl cellulose, polyvinylpyrrolidone, dextran, low-viscosity sodium alginate, Avicel PH 102 and lactose monohydrate. The results obtained revealed that the initial release of melatonin was more delayed from coated tablets than from the respective uncoated. The matrix tablets showed an initial fast release that followed a sustained mode, demonstrating that the use of various excipients results to different controlled-release behaviors [19].

Also, bilayered tablets incorporating an immediate-release layer and a sustained-release layer were developed by the same group of researchers using as excipients ethyl cellulose, dextran, low-viscosity sodium alginate, Avicel PH 102, lactose monohydrate, iron oxide pigment red 30 and magnesium stearate. The dissolution results revealed immediate and sustained drug release [19]. Another group of researchers also prepared bilayered tablets of MLT incorporating an immediate-release part consisting of MLT-*b*-cyclodextrin inclusion complex and a sustained-release part containing MLT in hydroxypropyl methylcellulose K15 M and Carbopol 971P. The results showed an initial burst followed by a near zero-order release pattern for a period of 8 h [20].

Furthermore, scientists have designed an MLT oral formulation to provide immediate and controlled release, which consisted of MLT-loaded sugar beads coated with 20% Aquacoat®. The in vivo results showed average peak plasma concentration at about 600 pg/ml that maintained at approximately 100 pg/ml over 8 h, indicating biphasic release [20].

Researchers have also utilized principles of nanotechnology to make micro−/nanoparticles containing MLT that could be further formulated to solid *per os* modified-release formulations. Thus, MLT was loaded in poly(*D*,*L*-lactide-co-glycolide) nanoparticles and microparticles (diameter of 200 nm and 3.5 mm, respectively). The cumulative release curves for nano- and microparticles revealed that for PLGA nano-10 and PLGA nano-20, approximately 30 and 20% of melatonin were released, respectively, within the first 24 h, as due to the diffusion of melatonin molecules located closer to the particle surface. At the end of 40 days, approximately 65% of the loaded melatonin was released from PLGA nanoparticles by diffusion mechanism [21]. Similarly, MLT was encapsulated into poly(lactic-co-glycolic acid) microspheres, and the release results indicated a dual pattern: a low initial burst release (around 40%) after the first 3 days and a relatively prolonged release over 25 days (around 85% of total MLT release) [22]. Furthermore, scientists have prepared MLT nanocapsules with Eudragit® S100. This formulation revealed a modified-release profile, which when fitted to a monoexponential model revealed that the MLT release mechanism was controlled by swelling and dissolution of the polymer [23].

3. Employment of DoE for the development of novel MLT modified-release formulations

In the vast majority of experimental procedures in all scientific fields, the optimal conditions are reached by modifying the levels of one factor at a time (OFAT) while keeping all the rest that seem to affect the response constant.

This classical strategy usually requires a large number of experimental runs and subsequent working hours. However, it ignores any potential interactions among the factors, and this is a major drawback, as it may result in an eventually ineffective procedure. On the contrary, a more organized way of conducting experiments, based on statistics, could be cost-effective and time-saving and also enable reaching the real optimal conditions. There are various chemometric approaches, but the most suitable for optimization of a procedure is the design of experiments.

DoE has been applied in many fields, including pharmaceutical product development [24–28]. It is gaining an increasing interest among pharmaceutical researchers, as more and more are becoming familiar with this approach, due to the relatively recent requirement for quality-by-design (QbD) principles. Furthermore, DoE has proven its usefulness in a variety of pharmaceutical applications in this field. Its major advantage has to do with obtaining the optimal conditions among factors for the desired values of responses by conducting a small number of experiments. That way DoE can resolve problems in complex systems, which cannot be easily managed by the trial-and-error approach.

Among the various types of designs like (fractional) factorial, Box–Behnken, central composite design (CCD), etc., the D-optimal design has been established as a robust design strategy. It enables the assessment of both numerical and categorical factors [29], and regarding numerical factors, the latter are examined at many different levels (design matrix), and not at 3–5, as the more classical designs. These levels are generated automatically by computer algorithms from relevant softwares, in order to satisfy the D-optimality criterion, aiming to minimize the generalized variance of the estimated regression coefficients without increasing the total number of experimental runs.

Such design was employed in the study presented by Vlachou et al. [17] regarding MLT controlled-release matrix formulations. One categorical factor, namely, the M.W. of polyvinylpyrrolidone (PVP), was chosen (M.W.: 10.000, low, and 55.000, high), and two numerical factors, namely, the mass (mg) of PVP and hydroxypropyl methylcellulose K15 M, were selected. When a modified release is the aim, as in the current study, setting the right responses is very critical. Herein, the need for a fully release melatonin in a controlled manner within 8 h was the reason for setting as responses the time for 50% drug dissolution at pH = 1.2 and the diffusional exponent (*n*) at pH values 1.2 and 7.4. Initially, a quick melatonin's release is needed for treating sleep-onset problems, while its subsequent slow release is needed to improve sleep quality and/or to assist maintain sleep. Therefore, T50% (pH: 1.2) should be ≤150 min, so that an initial dose will be released to aid the sleep onset of patients, and n (pH: 1.2) = 0.89, in order to achieve zero-order release kinetics and Case II diffusion, and n (pH: 7.4) = 0.80 for first-order release kinetics and anomalous diffusion.

The experiments were conducted as suggested by the experimental plan of Design-Expert software, and then suitable quadratic models were obtained for all (3) responses, satisfying all statistical criteria (ANOVA test, lack-of-fit test, R^2, *adj.* R^2 and *pred.* R^2 values). The next step was to estimate the overall optimal conditions, and thus Derringer's desirability function was employed [30], taking into account the necessity for simultaneous optimization of the aforementioned objectives/responses. Desirability function is a tool that is usually included in experimental design softwares and therefore very useful for projects in pharmaceutical development. Each predicted response $\hat{Y}i$ and experimentally obtained response Yi can be transformed to a desirability function di. The latter can have a value from 0 to 1, where $di = 0$ represents completely undesirable response and $di = 1$ represents completely desirable or ideal response. The individual desirability scores di can then be combined on a single overall (global) desirability D, which is optimized to find the optimum set of input variables:

$$D = (d_1 \times d_2 \times \ldots \times d_n)^{\frac{1}{n}} \quad (1)$$

with n denoting the number of responses.

In order to reach to optimal solution, the importance of responses should be set by adjusting the importance coefficients. T50% (pH, 1.2) was set as the most important response for consideration, while the rest two were of equal importance. Furthermore, weights (which denote the desired trend of the response within itself) and the range of responses could be changed, according to defined objectives. The optimal solution was reached with a value of global desirability of 0.907, which can be considered as very satisfactory. The suggested solution was performed, and the obtained results were in agreement with the goals defined for the responses and the predictions of the software. Consequently, with just 17 experiments defined by Design-Expert software and few preliminary in order to set the limits of the factors for the experimental plan, a novel MLT modified-release oral solid dosage form was developed.

To the best of our knowledge, the previous study was the only attempt to develop novel and improved MLT formulations by utilizing DoE. There has been a previous study [31] in which a different chemometric tool, artificial neural networks (ANN), was utilized. In that study, researchers prepared 27 different tablet formulations with different amounts of hydroxypropyl methylcellulose, xanthan gum and Carbopol® 974P NF. These formulations were subjected to drug release studies, using dissolution test data as inputs for ANN. The authors suggest that ANN with nine neurons in the hidden layer had the best results, meaning that it could predict, after training, dissolution data. In other words, this was a completely different strategy, based on training of the network and prediction of response values for novel (but not very different from training data) excipient mixtures. The optimal solution may not be reached, as the new mixtures are suggested by the user (trial and error) and then tested by the software. On the contrary, DoE is a tool for optimization of a procedure and not prediction of response values, and therefore it is recommended for application in such projects.

4. Conclusions

This analysis aims at the review of the latest advances of MLT modified-release oral solid dosage forms including the design of experiments approach. Many scientists have focused on the different ways in manufacturing modified-release oral solid formulations by using various excipients, dosage forms (multilayer or bilayer, coated or uncoated tablets), liposomes, alginate beads, nanofibre mats and nano−/microparticles, or by employing a variety of techniques (i.e. dry coating, electrospinning, experimental design, etc.) in order to gain knowledge for the production of such dosage forms.

Conflict of interest

The authors declare no conflict of interest.

Author details

Angeliki Siamidi[1*] and Yannis Dotsikas[2]

1 School of Health Sciences, Department of Pharmacy, Division of Pharmaceutical Technology, National and Kapodistrian University of Athens, Athens, Greece

2 School of Health Sciences, Department of Pharmacy, Division of Pharmaceutical Chemistry, National and Kapodistrian University of Athens, Athens, Greece

*Address all correspondence to: asiamidi@pharm.uoa.gr

References

[1] Vlachou M, Eikosipentaki A, Xenogiorgis V. Pineal hormone melatonin: Solubilization studies in model aqueous gastrointestinal environments. Current Drug Delivery. 2006;**3**:255-265. DOI: 10.2174/156720106777731073

[2] Vlachou M, Tsiakoulia A, Eikosipentaki A. Controlled release of the pineal hormone melatonin from hydroxypropylmethylcellulose/sodium alginate matrices in aqueous media containing dioctyl sulfosuccinate. Current Drug Discovery Technologies. 2007;**4**:31-38. DOI: 10.2174/1570 16307781115449

[3] Vlachou M, Ioannidou V, Vertzoni M, Tsotinis A, Afroudakis P, Sugden D. Controlled release from solid pharmaceutical formulations of two nalkanoyl-4-methoxybicyclo[4.2.0] octa-1,3,5-trien-7-ethanamines with melatoninergic activity. Letters in Drug Design & Discovery. 2015;**12**:259-262. DOI: 10.2174/157018081166614102400 5226

[4] Zisapel N. Development of a melatonin-based formulation for the treatment of insomnia in the elderly. Drug Development Research. 2000;**50**(3-4):226-234. DOI: 10.1002/1098-2299(200007/08) 50:3/4<226::aid-ddr6>3.0.co;2-s

[5] Lyseng-Williamson KA. Melatonin prolonged release: In the treatment of insomnia in patients aged ≥55 years. Drugs and Aging. 2012;**29**(11):911-923. DOI: 10.1007/s40266-012-0018-z

[6] Vlachou M, Siamidi A. Chapter 9: Melatonin modified release formulations designed for sleep disorders. In: Drăgoi MC, editor. Melatonin - Molecular Biology, Clinical and Pharmaceutical Approaches. Rijeka. ISBN 978-1-78984-505-1: InTechOpen; 2018. DOI: 10.5772/intechopen.78337

[7] Vlachou M, Papamichael M, Siamidi A, Fragouli I, Afroudakis PA, Kompogennitaki R, et al. Comparative in vitro controlled release studies on the chronobiotic hormone melatonin from cyclodextrins-containing matrices and cyclodextrin: Melatonin complexes. International Journal of Molecular Sciences. 2017;**18**:1641. DOI: 10.3390/ ijms18081641

[8] Vlachou M, Tragou K, Siamidi A, Kikionis S, Chatzianagnostou AL, Mitsopoulos A, et al. Modified in vitro release of the chronobiotic hormone melatonin from matrix tablets based on the marine sulfated polysaccharide ulvan. Journal of Drug Delivery Science and Technology. 2018;**44**:41-48. DOI: 10.1016/j.jddst.2017.11.019

[9] Vlachou M, Stavrou G, Siamidi A, Flitouri S, Ioannidou V, Mavrokordopoulos S. N-acetylserotonin vs melatonin. In vitro controlled release from hydrophilic matrix tablets. Letters in Drug Design & Discovery. 2018;**15**:347-352. DOI: 10.2174/15701808 15666180404125519

[10] Zampakola A, Siamidi A, Pippa N, Demetzos C, Vlachou M. Chronobiotic hormone melatonin: Comparative in vitro release studies from matrix tablets and liposomal formulations. Letters in Drug Design & Discovery. 2017;**14**(4):476-480. DOI: 10.2174/15701 808136661610061622 46

[11] Vlachou M, Kikionis S, Siamidi A, Tragou K, Ioannou E, Roussis V, et al. Modified in vitro release of melatonin loaded in nanofibrous electrospun mats incorporated into mono-layered and three-layered tablets. Journal of Pharmaceutical Sciences. 2019;**108**(2):970-976. DOI: 10.1016/j. xphs.2018.09.035

[12] Vlachou M, Kikionis S, Siamidi A, Tragou K, Kapoti S, Ioannou E, et al.

Fabrication and characterization of electrospun nanofibers for the modified release of the chronobiotic hormone melatonin. Current Drug Delivery. 2019;**16**(1):79-85. DOI: 10.2174/1567201815666180914095701

[13] Vlachou M, Siamidi A, Goula E, Georgas P, Pippa N, Sentoukas T, et al. Probing the release of the chronobiotic hormone melatonin from hybrid calcium alginate hydrogel beads. Acta Pharmaceutica. 2020:**70**. DOI: 10.2478/acph-2020-0037

[14] Kumar A, Agarwal SP, Khanna R. Modified release bi-layered tablet of melatonin using b-cyclodextrin. Die Pharmazie. 2003;**58**(9):642-644

[15] Lee BJ, Ryu SG, Cui JH. Formulation and release characteristics of hydroxypropyl methylcellulose matrix tablet containing melatonin. Drug Development and Industrial Pharmacy. 1999;**25**(4):493-501. DOI: 10.1081/DDC-100102199

[16] Priano L, Esposti D, Esposti R, Castagna G, De Medici C, Fraschini F, et al. Solid lipid nanoparticles incorporating melatonin as new model for sustained oral and transdermal delivery systems. Journal of Nanoscience and Nanotechnology. 2007;7:3596-3601. DOI: 10.1166/jnn.2007.809

[17] Vlachou M, Siamidi A, Konstantinidou S, Dotsikas Y. Optimization of controlled release matrix formulation of melatonin via experimental design. The Journal of Pharmaceutics and Drug Delivery Research. 2016;**5**(6):1-5. DOI: 10.4172/2325-9604.1000159

[18] Lee BJ, Min GH. Oral controlled release of melatonin using polymer-reinforced and coated alginate beads. International Journal of Pharmaceutics. 1996;**144**(1):37-46. DOI: 10.1016/S0378-5173(96)04723-0

[19] Vlachou M, Siamidi A, Pareli I, Zampakola A, Konstantinidou S. An account of modified release of melatonin from mono- and bi-layer solid pharmaceutical formulations. Journal of Pharmaceutical Sciences. 2016;**1**(4):10-14. DOI: 10.24218/vjpps.2016.19

[20] Lee BJ, Parrott KA, Ayres JW, Sack RL. Design and evaluation of an oral controlled release delivery system for melatonin in human subjects. The International Journal of Pharmaceutics. 1995;**124**(1):119-127. DOI: 10.1016/0378-5173(95)00088-Z

[21] Altındal DÇ, Gümüşderelioğlu M. Melatonin releasing PLGA micro/nanoparticles and their effect on osteosarcoma cells. Journal of Microencapsulation. 2015;**33**(1):53-63. DOI: 10.3109/02652048.2015.1115901

[22] Zhang L, Zhang J, Ling Y, Chen C, Liang A, Peng Y, et al. Sustained release of melatonin from poly (lactic-co-glycolic acid) (PLGA) microspheres to induce osteogenesis of human mesenchymal stem cells in vitro. Journal of Pineal Research. 2012;**54**(1):24-32. DOI: 10.1111/j.1600-079x.2012.01016.x

[23] Schaffazick SR, Pohlmann AR, Mezzalira G, Guterres SS. Development of nanocapsule suspensions and nanocapsule spray-dried powders containing melatonin. Journal of the Brazilian Chemical Society. 2006;**17**(3):562-569. DOI: 10.1590/s0103-50532006000300020

[24] Vatsaraj N, Zia H, Needham T. Formulation optimization of a sustained-release tablet of ketorolac tromethamine. Drug Delivery. 2002;**9**(3):153-159. DOI: 10.1080/10426500290095656

[25] Choi du H, Shin S, Khoa Viet Truong N, Jeong SH. A new experimental design method to optimize formulations focusing on a

lubricant for hydrophilic matrix tablets. Drug Development and Industrial Pharmacy. 2012;**38**(9):1117-1127. DOI: 10.3109/03639045.2011.641563

[26] Furlanetto S, Cirri M, Maestrelli F, Corti G, Mura P. Study of formulation variables influencing the drug release rate from matrix tablets by experimental design. European Journal of Pharmaceutics and Biopharmaceutics. 2006;**62**(1):77-84. DOI: 10.1016/j.ejpb.2005.07.001

[27] Martinello T, Kaneko TM, Velasco MV, Taqueda ME, Consiglieri VO. Optimization of poorly compactable drug tablets manufactured by direct compression using the mixture experimental design. International Journal of Pharmaceutics. 2006;**322**(1-2):87-95. DOI: 10.1016/j.ijpharm.2006.05.034

[28] Dey S, Mahanti B, Khila S, Mazumder B, Gupta SD. Formulation development and optimization of bilayer tablets of aceclofenac. Expert Opinion on Drug Delivery. 2012;**9**(9):1041-1050. DOI: 10.1517/17425247.2012.707187

[29] de Aguiar PF, Bourguignon B, Khots MS, Massart DL, Phan-Than-Luu R. D-optimal designs. Chemometrics and Intelligent Laboratory Systems. 1995;**30**(2):199-210. DOI: 10.1016/0169-7439(94)00076-X

[30] Malenović A, Dotsikas Y, Mašković M, Jančić-Stojanović B, Ivanović D, Medenica M. Desirability-based optimization and its sensitivity analysis for the perindopril and its impurities analysis in a microemulsion LC system. Microchemical Journal. 2011;**99**(2):454-460. DOI: 10.1016/j.microc.2011.06.022

[31] Martarelli D, Casettari L, Shalaby KS, Soliman ME, Cespi M, Bonacucina G, et al. Optimization of melatonin dissolution from extended release matrices using artificial neural networking. Current Drug Delivery. 2016;**13**(4):565-573. DOI: 10.2174/1567201812666150608101528

Chapter 6

The Release Kinetics of Melatonin from Innovative Dosage Forms: The Role of the Fractal Geometry of the "Vehicle"

Natassa Pippa and Costas Demetzos

Abstract

Melatonin (N-acetyl-5-methoxytryptamine) is an antioxidant active pharmaceutical ingredient with numerous applications as medicine and nutraceutical. Melatonin, a hormone synthesized by the pineal gland, has a significant role in the regulation of the circadian biological clock. The aim of this chapter is to present the conventional solid and liquid forms (i.e., tables, capsules, suspensions, etc.) and the nanoformulations (i.e., liposomes, niosomes, polymeric nanoparticles, chitosomes, calcium alginate beads, etc.) of melatonin and to give special attention to its release kinetics from the pharmaceutical vehicle. These systems have been designed and developed as platforms for the delivery and release of melatonin. In all cases, the controlled release of melatonin is the main goal of its loading into drug delivery platforms. Fractal analysis is a mathematical tool to quantify nature and physical systems' complexity. These systems have been characterized as fractal objects, due to their fractional dimensions. In this chapter, we are probing the interrelationship between the fractal dimension of pharmaceutical vehicle and the release profile of melatonin. Several examples will be given in order to understand in depth the reason of controlled-release profile of melatonin and its added value for the development of a new medicine and/or nutraceutical.

Keywords: nanosystems, drug delivery, kinetics, release, fractals

1. Introduction

According to the Drug and Lactation Database, "Melatonin is the hormone produced by the pineal gland that plays a role in regulating sleep and circadian rhythm as well as a possible role in gut-brain signaling." As it is stated in the Drug and Lactation Database, melatonin (methoxyindole) is used for the organization of the circadian rhythms, especially core temperature and sleep–wake rhythms [1]. Melatonin is also characterized as a full-service anticancer agent due to its functions: inhibition of initiation, progression, and metastasis phases of tumors [2]. Aside from its antioxidant, anticancer, antitumor, anti-inflammatory, antiaging, antidiabetic, antiviral, and neuroprotective activities, melatonin exhibits a therapeutic potential in the treatment of asthma, respiratory diseases or infections, chronic obstructive pulmonary disease, lung cancer, pleural cavity diseases, as well as vascular pulmonary diseases [3]. Melatonin is also used as a food supplement and nutraceutical. The dosage and release profile of melatonin are very crucial factors that affect the effectiveness of treatment, especially in older adults [4]. According to a recent critical analysis, in older adults, the use of the lowest possible dose of immediate-release formulation of melatonin is appropriate to best mimic the normal physiological circadian rhythm of melatonin and to avoid prolonged, supra-physiological blood levels [4].

Melatonin has been encapsulated in different conventional and nanotechnological systems [5]. In the majority of the cases, the aim of the incorporation of melatonin into formulations is to achieve controlled or sustained release. The aim of this chapter is to present the conventional solid and liquid forms (i.e., tables, emulsions, suspensions, etc.) and the nanoformulations of melatonin and to give special attention to release kinetics from the pharmaceutical vehicle.

Furthermore, fractal analysis is a mathematical tool to quantify nature and physical systems' complexity [6]. Fractals have been observed in powdered drug substances, in excipients, and in their mixtures, as well as in semifluid dosage forms like gels and emulsions [6]. Fractals have been used to describe the dimensions of dosage forms, such as tablets, matrix tablets, and spheres [6]. The application of fractal geometry for the quantification of the dimensionality of advanced drug delivery nanosystems (aDDnS) recently appeared in the literature [5, 6]. For example, liposomes, micelles, polymersomes, and other nanosystems are fractal objects [6]. Additionally, the fractal and fractional kinetics can model very close to the reality the release of drugs from polymeric matrices and other dosage forms, both solid and liquid [7–9].

In this chapter, we are going to find the interrelationship between the fractal dimension of pharmaceutical vehicle and the release profile of melatonin. Several examples will be given in order to understand in depth the reason of controlled-release profile of melatonin and its added value for the development of a new medicine and/or nutraceutical.

2. Dosage forms of melatonin

2.1 Conventional dosage forms of melatonin

Reiter et al. summarized what is known about the function of melatonin in the oral cavity [10]. Melatonin is released into the saliva by the acinar cells of the major salivary glands and via the gingival fluid [10]. Functions of melatonin in the oral cavity are likely to relate primarily to antioxidant activities [10]. A case series study revealed that the light level and duration of exposure determine the impact of self-luminous tablet on melatonin suppression [11]. Hydrophilic polymer matrices composed of hydroxypropyl methylcellulose, xanthan gum, and Carbopol®974P NF in different amounts were formulated in tablet forms [12]. These tablets exhibited a prolonged-release profile of melatonin [12]. Monolayered and three-layered tablets, incorporating nanofibrous mats composed of cellulose acetate and polyvinylpyrrolidone loaded with MLT, were prepared and exhibited a prolonged-release profile of melatonin, too [13]. The release profile of Circadin® tablets is presented in the recent literature. Circadin® is a prolonged-release tablet, the only licensed melatonin formulation available in the UK [14]. According to Chua et al., the division of tablet into two or four halves did not affect the prolonged-release characteristics [14]. We can observe this kinetics in **Figure 1**. Furthermore, immediate-release tablets are available in Greek markets as food supplements. This formulation investigation is a composition of natural ingredients, which have a relaxant,

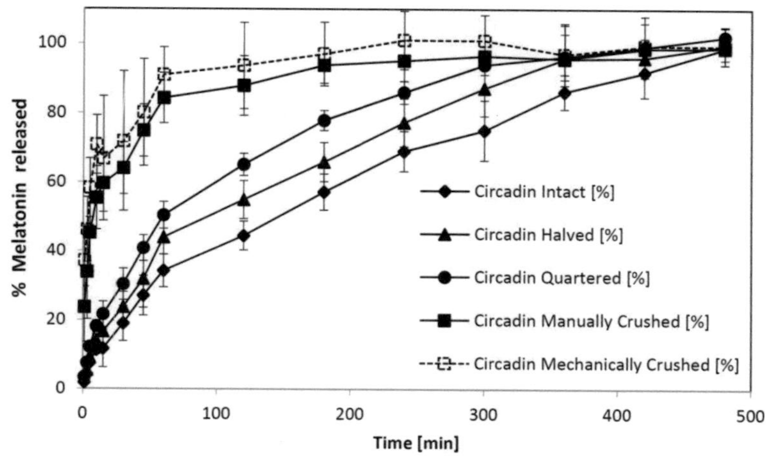

Figure 1.
Comparison of dissolution profiles for the release of melatonin from Circadin tablets in an intact, halved, quartered, and crushed form (data represented mean ± SD, n = 6) (adapted from [14]). The division of tablet into two or four halves did not affect the prolonged-release characteristics.

anxiolytic, and sleep-inducing action. The active ingredients of the formulations are melatonin, L-tryptophan (an important amino acid, which is converted in the brain to serotonin (neurotransmitter that contributes to a normal sleep cycle; serotonin is sequentially converted to melatonin)), passiflora, valeriana, hawthorn, and eschscholtzia that increase GABA levels in the brain. For an immediate-release formulation, the amount of drug released should not be less than 80% of the labeled amount at 30 minutes [15]. This release profile of melatonin is suitable for alleviating certain insomnia-related problems and particularly those arising from sleep-onset difficulties [15].

Other conventional dosage forms that are used for the melatonin delivery are soft capsule gels [16, 17]. Soft gel capsules improved the bioavailability of melatonin in humans even when the administered dose was reduced [16]. Considering the number of conditions in which melatonin supplementation is recommended, this evidence could support a broader use of melatonin in clinical practice, especially in the field of nutraceuticals [16]. Sublingual solution and hard capsules of melatonin have been also appeared as dosage forms in the literature [17]. The sublingual solution was prepared with glycerin, ethyl alcohol, stevia powder extract, and tutti-frutti flavor. The concentration of melatonin was equal to 10 mg/ml [17]. The hard capsules are composed of Methocel E4M and lactose anhydrous, and the amount of melatonin was 3 mg per capsule. Both of the aforementioned formulations were found to be stable in accelerated conditions [17].

2.2 Advanced drug delivery nanosystems of melatonin

2.2.1 Liposomes and lipid drug delivery nanosystems

Liposomes are bilayers composed of phospholipids [6]. They are used as drug and vaccine delivery systems and as cellular membrane models [6]. They are biocompatible and biodegradable [6]. The location of melatonin in 1,2-dioleoyl-sn-glycero-3-phosphocholine (DOPC) and 1,2-dipalmitoyl-sn-glycero-3-phosphocholine (DPPC) model membranes was investigated by several techniques such as small-angle neutron scattering (SANS) and molecular dynamics (MD) simulations [18]. The location of melatonin in those lipid membranes is illustrated in **Figure 2** [18]. The interactions of melatonin with cellular membrane models can be a road map to elucidate the protective effect of melatonin against the formation of amyloid-beta (Aβ) proteins of Alzheimer's disease [19]. In **Figure 2**, we can also observe that melatonin is located in the interface between head groups and lipid tails, while cholesterol is located parallel to lipid chains [18]. The "stabilizing" effect of melatonin, a naturally occurring hormone produced by the brain's pineal gland, on phase-separated model membranes mimicking the outer leaflet of plasma membranes was also investigated [20]. For example, melatonin stabilizes the liquid-ordered/liquid-disordered phase coexistence over an extended range of temperatures. Melatonin appeared to induce re-ordering effects in liposome and Langmuir monolayers [20].

Melatonin-loaded liposomes (MLL) were successfully prepared using rapid expansion of supercritical solution technology [21]. The system is composed of phosphatidylcholine-cholesterol-melatonin at 20:2:1 molar ratio, and the size of the liposomes was found to be around 100 nm [21]. The release kinetics of melatonin shows slow-release features in early digestive stages and more through characteristics in later stages of simulated gastric fluids [21]. Furthermore, vesicular (liposomal and nanoencapsulated) forms of melatonin efficiently downregulate sodium fluoride-induced rat hepato- and broncho-TNF-α, TGF-β expressions, and associated oxidative injury, as well as oxidative damage in sodium fluoride (NaF)-treated lungs and liver [22]. The nanoencapsulated melatonin was evaluated as a more powerful remedial therapy in comparison with liposomes, in terms of its efficacy in regulating NaF-intoxicated oxidative injury [22]. Melatonin also encapsulated into liposomes produced using supercritical carbon dioxide (an easier technique compared to thin-film hydration method) [23]. The release profile of melatonin is more or less the same for the liposomes prepared by the two techniques (i.e., supercritical carbon dioxide and thin-film hydration method), but exhibits differences compared to tablets [23].

The hypotensive melatonin analogue 5-methoxycarbonylamino-N-acetyltryptamine (5-MCA-NAT) was loaded into hybrid liposomes (i.e., polymer-grafted liposomes). These hybrid formulations were combined with mucoadhesive (sodium hyaluronate or carboxymethylcellulose) or amphiphilic block thermosensitive (poloxamer) polymers to prolong the release profile of the active ingredient. The prepared hybrid liposomes' size was found to be between 150 to 200 nm with low polydispersity and zeta potential near zero. The release profile of the 5-MCA-NAT melatonin was found to be dependent on the nature of the polymeric guest. In the presence of the polymer, the released ratio of melatonin was also decreased

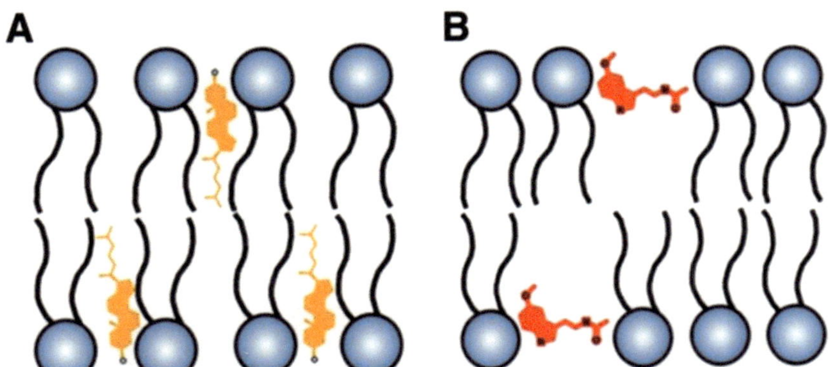

Figure 2.
Schematics illustrating the proposed locations of cholesterol and melatonin in the lipid membrane: (A) cholesterol, (B) melatonin. (adapted from [18]).

in comparison with the pure liposomes. The hypotensive effect of the prepared systems was further investigated in rabbit eyes.

Melatonin was also incorporated into chitosomes [24]. Chitosomes are chitosan-coated liposomes that represent an alternative to conventional liposomes since they present better stability and bioadhesivity [24]. Chitosomes are prepared by using a different molar ratio of the active ingredient and the polymer. In all cases, the amount of the phosphatidylcholine was constant [24]. These formulations exhibit size between 200 and 250 nm, with negative zeta potential and encapsulation efficiency between 30 and 60% [24]. The amount of chitosan exhibited the key role for the stability of the polymer-coated liposomes and their loading properties [24].

Last but not least, melatonin and its structural analogues do not possess antioxidant properties on Fe(2+)-initiated peroxidation of sonicated liposomes made of retinal lipids [25]. The in vitro protective effects of melatonin against oxidation of 1-palmitoyl-2-linoleoyl-sn-glycero-3-phosphocholine (PLPC) liposomes and low-density lipoproteins (LDL; 3 g/L total concentration) by hydroxyl radicals produced by water gamma radiolysis were also investigated [26].

Solid lipid microparticles were designed as an oral pulsatile system for the delivery of melatonin to pediatric patients [27].

2.2.2 Niosomes

Niosomes are composed of surfactants (Tweens and Spans) and are vesicular drug delivery systems. The unique structure of niosome presents an effective novel drug delivery system with the ability of loading both hydrophilic and lipophilic drugs [28, 29]. A transmucosal niosome gel was developed to improve the pharmacokinetics of exogenous melatonin [30]. The melatonin niosome gel was characterized by several physicochemical techniques, and melatonin levels were determined in healthy volunteers [30]. Oral transmucosal melatonin niosome gels, at different molar ratio of the active ingredient, topically applied in 14 healthy volunteers in a randomized double-blinded crossover design with a 7-day washout, gave dose-proportional pharmacokinetics, with improved absorption and prolonged systemic circulation [30]. Additionally, melatonin-loaded elastic niosomes were prepared and lyophilized [31]. The lyophilized niosomal system was incorporated into the Pickering emulsion. Ex vivo permeation studies revealed 58% of melatonin were permeated through the rat skin while 37% of melatonin accumulated in the skin after 24 hours. This formulation is effective for UV-induced skin damage [31].

2.2.3 Polymeric nanoparticles and polymer-based drug delivery nanosystems

Polymeric nanoparticles include a variety of systems composed of polymers. The polymers should be biocompatible for drug delivery applications. Generally, polymers exhibit nice drug loading and release properties. Some polymers like Pluronics® are FDA approved. Pohlmann et al. (2010) have shown that the in vitro antioxidant effect of melatonin against lipid peroxidation in microsomes and liposomes can be improved by encapsulation of the antioxidant drug in polymeric nanoparticles. Polymeric nanoparticles (nanocapsules or nanospheres) have been used to improve the melatonin efficacy and release, too [32]. For example, incorporation in polymeric nanocapsules improves the antioxidant effect of melatonin against lipid peroxidation in mice brain and liver. It should be pointed out that the melatonin-loaded polysorbate 80-coated nanocapsules caused a marked reduction on lipid peroxidation levels in all studied tissues and increased the total antioxidant reactivity in the hippocampus [32].

Two types of polymeric nanoparticles have been also investigated for ocular administration of melatonin [33]. For example, lecithin/chitosan nanoparticles with size around 250 nm and Pluronics® F127/chitosan micelles with size around 20 nm have been designed and developed in order to improve the bioavailability of melatonin in the eyes [34]. According to the authors, "the permeability study results confirmed the permeation enhancing effect of F127, which was hindered in the presence of chitosan. Lecithin/chitosan nanoparticles were characterized by prominent mucoadhesive properties and prolonged melatonin release, which was shown to control melatonin permeation across an in vitro corneal epithelial model. Such properties demonstrate the potential for nanoparticles to provide an extended pre-corneal residence time of melatonin, ensuring higher eye-related bioavailability and extended intraocular pressure reduction compared to melatonin in both aqueous and micelle solutions" [34].

Poly(D,L-lactide-co-glycolide) (PLGA) polymers have been used for the preparation of nanoparticles and microparticles loaded with melatonin [35, 36]. Both of them are prepared by the emulsion-diffusion-evaporation method and the addition of 0.2% (w/v) melatonin in the aqueous phase. The size of nanoparticles was about 200 nm and the entrapment efficiency around 14%, while the size of microparticles was about 3.5 micrometers and the encapsulation efficiency 27% [35, 36]. The toxicity and the effectiveness of the prepared systems are also evaluated using cell lines. According to the results, melatonin could be an adjunct to the routine chemotherapy of osteosarcoma by encapsulating it into PLGA polymeric delivery platform [36]. Additionally, PLGA nanoparticles and polysorbate 80-coated PLGA nanoparticles (PLGA-PS80) increase the in vitro antioxidant activity of melatonin [36]. The sizes of the PLGA-PS80 and PLGA nanoparticles were 212 and 187 nm, and the encapsulation entrapment of melatonin was 26 and 41%, respectively [37]. The release kinetics of melatonin followed the second-order model during studies from nanoparticles, while PLGA-PS80 presented more prolonged melatonin release [37]. The spherical shape of nanoparticles and the strong interactions due to negative zeta potential are the possible explanation of the release kinetics of melatonin from nanoparticles [37].

Furthermore, melatonin was loaded into chitosan-tripolyphosphate nanoparticles [38]. Melatonin nanoparticles protect against etoposide-induced genotoxicity in the HepG2 cell line (etoposide is one of the most effective chemotherapeutic agents used in the treatment of various types of tumors) [38]. An increased nose-to-brain delivery of melatonin mediated by polycaprolactone nanoparticles for the treatment of glioblastoma has been also designed and developed [39]. No cytotoxic effect was observed against non-tumor cells [39]. Another interesting formulation of melatonin is hybrid hydrogels composed of calcium alginate beads and combinations of polymers such as polyvinylpyrrolidone (PVP_{10000} and PVP_{55000}), hydroxypropyl methylcellulose ($HPMC_{15000}$ and $HPMC_{100000}$) at different molar weights, lactose monohydrate, and as a surfactant sodium laureth sulfate (SLS) [40]. In all cases, the encapsulation efficiency of melatonin was very high, around 80% [40]. The swelling studies and the release profile were found to be dependent on the presence of the polymer [40]. Fickian diffusion mechanism and burst release were also observed [40]. The nature (architecture and molecular weight) of the polymeric guest altered the physicochemical behavior of the calcium alginate beads and the release of melatonin, too [40].

2.3 Release kinetics of melatonin from delivery platforms

Melatonin has been encapsulated in different delivery carriers, as mentioned above. The majority of these carriers are summarized in **Table 1**. We should highlight

Conventional forms	Advanced drug delivery systems
Tablets	Liposomes/lipid carriers
• Polymeric matrices	• Chitosan-coated liposomes (chitosomes)
• Three-layered tablets	• Polymer-grafted liposomes
Sublingual suspensions	• Solid lipid microparticles as pulsatile system
Capsules	Niosomes
• Soft gels	Polymeric nanoparticles
• Hard gels	• PLGA nanoparticles and microparticles
	• Polysorbate 80-coated PLGA nanoparticles

Table 1.
The delivery systems of melatonin.

that the reason of the design and the development of different formulations of melatonin is the achievement of its controlled/programmed release. There are systems of immediate release of melatonin, i.e., tablets and some others where the release of melatonin is prolonged, i.e., polymeric nanoparticles. The treatments of sleep-onset problems and/or sleep maintenance discomforts are the goals of preparing a delivery platform of melatonin as medicine or as food supplement/nutraceutical. Recently, a comparative study of the in vitro release of melatonin from matrix tablets and liposomal formulation appeared in the literature [41]. The matrix tablets used were comprised of HPMC and dextran. Moreover, melatonin was encapsulated into conventional liposomes composed of DPPC and dipalmitoyl-phosphatidyl glycerol (DPPG), in an attempt to compare the hormone's release profile from liposomal formulations with its respective release from matrix tablets [41]. Some of the formulations of the matrix tablets and the liposomes exhibit the same release behavior ideal for the maintenance of sleep [41]. On the other hand, the burst release of melatonin from some other matrix tablet formulations is ideal for the fast sleep onset [41].

In order to design and develop an ideal drug delivery platform combining a burst release of melatonin accompanied by a prolonged release, mathematical modeling and simulations are needed. In that case, the fractal nature of the formulation should be taken into consideration [6–8]. The last one is very important because the fractal dimensionality of these systems is closer to their real dimensions [6–8]. Both conventional systems like tablets and capsules and nanosystems like liposomes and polymeric nanoparticles are fractal objects. In other words, the mechanistic explanation of the release profile of melatonin should be based on the fractal dimensions of the drug delivery systems.

3. Conclusions

Melatonin is a chronobiotic hormone used for the treatment of sleep problems and disorders. In this chapter, we presented the formulations of melatonin that appeared in the literature. Tablets, capsules (hard and soft gels), and the advanced drug delivery systems of pharmaceutical nanotechnology offer the possibility of encapsulation efficiency of melatonin and ideal release properties. Several routes of administration have been proposed, but in the majority of the cases, the per os administration is the most popular route of administration, especially of the nutraceuticals. The interrelationship between the fractal dimension of pharmaceutical vehicle and the release profile of melatonin is the key point for the development of delivery platforms of melatonin as medicines and food supplements/nutraceuticals.

Conflict of interest

The authors confirm that this chapter content has no conflict of interest.

Author details

Natassa Pippa* and Costas Demetzos
Division of Pharmaceutical Technology, Department of Pharmacy, School of Health Sciences, National and Kapodistrian University of Athens, Athens, Greece

*Address all correspondence to: natpippa@pharm.uoa.gr

References

[1] Claustrat B, Leston J. Melatonin: Physiological effects in humans. Neurichirugie. 2015;**61**:77-85. DOI: 10.1016/j.neuchi.2015.03.002

[2] Reiter RJ, Rosales-Corral SA, Tan DX, Acuna-Castroviejo D, Qin L, Yang SF, et al. Melatonin, a full service anti-cancer agent: Inhibition of initiation, progression and metastasis. International Journal of Molecular Sciences. 2017;**843**:1-47. DOI: 10.3390/ijms18040843

[3] Habtemariam S, Daglia M, Sureda A, Selamoglu Z, Gulhan MF, Nabavi SM. Melatonin and respiratory diseases: A review. Current Topics in Medicinal Chemistry. 2017;**17**:467-488

[4] Vural EM, van Munster BC, de Roiiij SE. Optimal dosage for melatonin supplementation therapy in older adults: A systematic review of current literature. Drugs & Aging. 2014;**31**:441-451. DOI: 10.1007/s40266-014-0178-0

[5] Demetzos C, Pippa N. Advanced drug delivery nanosystems (aDDnSs): A mini review. Drug Delivery. 2014;**21**:250-257. DOI: 10.3109/10717544.2013.844745

[6] Pippa N, Dokoumetzidis A, Demetzos C, Macheras P. On the ubiquitous presence of fractals and fractal concepts in pharmaceutical sciences: A review. International Journal of Pharmaceutics. 2013;**456**:340-352. DOI: 10.1016/j.ijpharm.2013.08.087

[7] Pippa N, Dokoumetzidis A, Pispas S, Demetzos C. The interplay between the rate of release from polymer grafted liposomes and their fractal morphology. International Journal of Pharmaceutics. 2014;**465**:63-69. DOI: 10.1016/j.ijpharm.2014.02.010

[8] Dokoumetzidis A, Macheras P. The changing face of the rate concept in biopharmaceutical sciences: From classical to fractal and finally to fractional. Pharmaceutical Research. 2011;**28**:1229-1232. DOI: 10.1007/s11095-011-0370-4

[9] Kosmidis K, Macheras P. On the dilemma of fractal or fraction kinetics in drug release studies: A comparison between Weibull and Mittag-Leffler functions. International Journal of Pharmaceutics. 2018;**543**:269-273. DOI: 10.1016/j.ijpharm.2018.03.060

[10] Reiter RJ, Rosales-Corall SA, Liu XY, Acuna-Castroviejo D, Escames G, Tan DX. Melatonin in the oral cavity: Physiological and pathological implications. Journal of Periodontal Research. 2015;**50**:9-17. DOI: 10.1111/jre.12176

[11] Wood B, Rea MS, Plitnick B, Figueiro MG. Light level and duration of exposure determine the impact of self-luminous tablets on melatonin suppression. Applied Ergonomics. 2013;**44**:237-240. DOI: 10.1016/j.apergo.2012.07.008

[12] Martarelli D, Casettari L, Shalby KS, Soliman ME, Cespo M, Bonacucina G, et al. Optimization of melatonin dissolution from extended release matrices using artificial neural networking. Current Drug Delivery. 2016;**13**:565-573

[13] Vlachou M, Kikionis S, Siamidi A, Tragou K, Ioannou E, Roussis V, et al. Modified in vitro release of melatonin loaded in nanofibrous electrospun mats incorporated into monolayered and three-layered tablets. Journal of Pharmaceutical Sciences. 2019;**108**:970-976. DOI: 10.1016/j.xphs.2018.09.035

[14] Chua HM, Hauet Richer N, Swedrowska M, Ingham S, Tomlin S, Forbes B. Dissolution of intact, divided and crushed Circadin tablets: Prolonged

vs. immediate release of melatonin. Pharmaceutics. 2016;**8**:1-11. DOI: 10.3390/pharmaceutics8010002

[15] Vlachou M, Siamidi A, Pippa N, Fikioris N. Fast sleep onset pharmaceutical formulation of the chronobiotic hormone melatonin. Advanced Science Engendering and Medicine. **12**:438-441. DOI: 10.1166/asem.2019.2542

[16] Proietti S, Carlomagno G, Dinicola S, Bizzarri M. Soft gel capsules improve melatonin's bioavailability in humans. Expert Opinion on Drug Metabolism & Toxicology. 2014;**10**:1193-1198. DOI: 10.1517/17425255.2014.943183

[17] Haywood A, Burrell A, Breda KV, George R, Testa C, Norris R. Stability of melatonin in an extemporaneously compounded sublingual solution and hard gelatin capsule. International Journal of Pharmaceutical Compounding. 2009;**13**:170-174

[18] Drolle E, Kuerka N, Hoopes MI, Choi Y, Katsara J, Karttunen M, et al. Effect of melatonin and cholesterol on the structure of DOPC and DPPC membranes. Biochimica et Biophysica Acta. 1828;**2013**:2247-2254

[19] De Lima VR, Caro MS, Munford ML, Desbat B, Dufourc E, Pasa AA, et al. Influence of melatonin on the order of phosphatidylcholine-based membranes. Journal of Pineal Research. 2010;**49**:169-175. DOI: 10.1111/j.1600-079X.2010.00782.x

[20] Bolmatov D, McClintic WT, Taylor G, Stanley CB, Do C, Collier CP, et al. Deciphering melatonin-stabilized phase separation in phospholipid bilayers. Langmuir. 2019;**35**:12236-12245. DOI: 10.1021/acs.langmuir.9b01534

[21] Zhang Q, Ou C, ye S, Song X, Luo S. Construction of nanoscale liposomes loaded with melatonin via supercritical fluid technology. Journal of Microencapsulation. 2017;**34**:687-698. DOI: 10.1080/02652048.2017.1376001

[22] Sana S, Ghosh S, Das N, Sarkar S, Mandal AK. Vesicular melatonin efficiently downregulates sodium fluoride-induced rat hepato- and broncho- TNAF-a, TGF -β expressions, and associated oxidative injury: A comparative study of liposomal and nanoencapsulated forms. International Journal of Nanomedicine. 2017;**12**:4059-4071

[23] Situ W, Song X, Luo S, Liang Y. A nano-delivery system for bioactive ingredients using supercritical carbon dioxide and its release behaviours. Food Chemistry. 2017;**228**:219-225. DOI: 10.1016/j.foodchem.2017.01.053

[24] Gonçalves MC, Mertins O, Pohlmann AR, Silveira NP, Guterres SS. Chitosan coated liposomes as an innovative nanocarrier for drugs. Journal of Biomedical Nanotechnology. 2012;**8**:240-250

[25] Fagali N, Catalá A. Melatonin and structural analogues do not possess antioxidant properties of Fe(2+)-initiated peroxidation of sonicated liposomes made of retinal lipids. Chemistry and Physics of Lipids. 2011;**164**:688-695. DOI: 10.1016/j.chemphyslip.2011.07.004

[26] Marchetti C, Sidahmed-Adrar B, Collin F, Jore D, Gardès-Albert M, Bonnefont-Rousselot D. Melatonin protects PLPC liposomes and LDL towards radical-induced oxidation. Journal of Pineal Research. 2011;**51**:286-296. DOI: 10.1111/j.1600-079X.2011.00889.x

[27] Albertini B, Di Savatino M, Melegari C, Passerini N. Formulating SLMs as oral pulsatile system for potential delivery of melatonin to pediatric population. International

Journal of Pharmaceutics. 2014;**469**:67-79

[28] Moghassemi S, Hadjizadeh A. Nano-niosomes as nanoscale drug delivery systems: An illustrated review. Journal of Controlled Release. 2014;**185**:22-36. DOI: 10.1016/j.jconrel.2014.04.015

[29] Abdelkader H, Alani AW, Alany RG. Recent advances in non-ionic surfactant vesicles (niosomes): Self-assembly, fabrication, characterization, drug delivery applications and limitations. Drug Delivery. 2014;**21**:87-100. DOI: 10.3109/10717544.2013.838077

[30] Priprem A, Nukulkit C, Johns NP, Laohasiriwong S, Yimtae K, Soontornpas C. Transmucosal delivery of melatonin-encapsulated Niosomes in a mucoadhesive gel. Therapeutic Delivery. 2018;**9**:343-357. DOI: 10.4155/tde-2018-0001

[31] Arslan Azizoglu G, Tuncay Tanriverdi S, Aydin Kose E, Ballar Kirmizibayrak P, Ozer O. Dual-prevention for UV-induced skin damage: Incorporation of melatonin-loaded elastic niosomes into octyl methoxycinnamate pickering emulsions. AAPS PharmSciTech. 2017;**18**:2987-2998. DOI: 10.1208/s12249-017-0786-1

[32] Pohlmann AR, Schaffazick SR, Creczynski-Pasa TB, Guterrs SS. Preparation of drug-loaded polymeric nanoparticles and evaluation of the antioxidant activity against lipid peroxidation. Methods in Molecular Biology. 2010;**610**:109-121. DOI: 10.1007/978-1-60327-029-8_7

[33] Schaffazick SR, Siqueira IR, Badejo AS, Jornada DS, Pohlmann AR, Netto CA, et al. Incorporation in polymeric nanocapsules improves the antioxidant effect of melatonin against lipid peroxidation in mice brain and liver. European Journal of Pharmaceutics and Biopharmaceutics. 2008;**69**:65-71. DOI: 10.1016/j.ejpb.2007.11.010

[34] Hafner A, Lovrić J, Romić MD, Juretić M, Pepić I, Cetina-Čižmek B, et al. Evaluation of cationic nanosystems with melatonin using an eye-related bioavailability prediction model. European Journal of Pharmaceutical Sciences. 2015;**75**:142-150. DOI: 10.1016/j.ejps.2015.04.003

[35] Milan AS, Campmany ACC, Naveros BC. Antioxidant nanoplatforms for dermal delivery: Melatonin. Current Drug Metabolism. 2017;**18**:437-453. DOI: 10.2174/1389200218666170222145908

[36] Altindal DÇ, Gümüşderelioğlu M. Melatonin releasing PLGA micro/nanoparticles and their effect on osteosarcoma cells. Journal of Microencapsulation. 2016;**33**:53-63. DOI: 10.3109/02652048.2015.1115901

[37] Martins LG, Khalil NM, Mainardes RM. PLGA nanoparticles and polysorbate-80- coated PLGA nanoparticles increase the In vitro antioxidant activity of melatonin. Current Drug Delivery. 2018;**15**:554-563. DOI: 10.2174/1567201814666170719112535

[38] Shokrzadeh M, Chassemi-Barghi N. Melatonin loading chitosan-tripolyphosphate nanoparticles: Application in attenuating etoposide-induced genotoxicity in HepG2 cells. Pharmacology. 2018;**102**:74-80. DOI: 10.1159/000489667

[39] de Oliveira Junior ER, Nascimento TL, Salomão MA, da Silva AGG, Valadares MC, Lima EM. Increase nose-to-brain delivery of melatonin mediated by polycaprolactone nanoparticles for the treatment of glioblastoma. Pharmaceutical Research. 2019;**36**:131. DOI: 10.1007/s11095-019-2662-z

[40] Vlachou M, Siamidi A, Goula E, Georgas P, Pippa N, Sentoukas T, et al. Probing the release of the chronobiotic hormone melatonin from hybrid calcium alginate hydrogel beads. Acta Pharmaceutica. 2020 In press

[41] Zampakola A, Siamidi A, Pippa N, Demetzos C. Chronobiotic hormone melatonin: Comparative in vitro release studies from matrix tablets and liposomal formulations. Letters in Drug Design & Discovery. 2017;**14**:476-480. DOI: 10.2174/157018081366616100616 2246

Section 5

Melatonin's Physiological Functions in Plants

Chapter 7

Review of Melatonin in Horticultural Crops

Yanyan Yan, Qinghua Shi and Biao Gong

Abstract

Melatonin is an indoleamine, abundant in animals and plants, which has the functions of regulating circadian rhythm, improving immunity and anti-aging in animals, and is a good health care product beneficial to human health. Recent studies have shown that melatonin has physiological functions including regulating plant growth, promoting seed germination, controlling root development and delaying leaf senescence. The antioxidant properties of melatonin give it the ability to strengthen plants' resistance to stress. The comprehensive researches in recent years, involving five aspects of "the biosynthetic pathway of melatonin in plants, the melatonin in horticultural crops and its influencing factors, the roles of melatonin in the growth and development of horticultural crops, in the response to stress of horticultural crops, the signal transduction network of melatonin in regulating plant growth and the development and stress resistance," are reviewed in the present paper. The application of melatonin in horticulture production is also discussed, which can provide a theoretical reference for the application of melatonin in horticultural production.

Keywords: development, growth, horticultural crops, melatonin, stress tolerance

1. Introduction

Melatonin, *N*-acetyl-5-methoxytryptamine, is an indole-like tryptamine. In 1958, Lerner et al. [1] extracted melatonin from the pineal gland of cattle and found that it was a kind hormone-like substance, which was widely involved in the growth and development regulation, as well as signal transduction, in animals. Before the 1990s, melatonin was recognized as an animal hormone. However, Balzer and Hardeland [2] successfully isolated and identified melatonin from single-celled algae, indicating the existence of natural melatonin in the plant kingdom. Since then, melatonin has become a research hotspot and received extensive attention from researchers in the field of plant and agricultural science. With the rapid development of material separation and identification technology, plant physiology and cell biology, molecular biology, sequencing technology and other research methods, scientists have conducted more comprehensive and in-depth studies on the synthesis pathway, concentration, distribution and biological function of melatonin in plants. Moreover, melatonin has a variety of regulatory effects in plants and has many benefits for human health. Therefore, the research on melatonin in horticultural crops is fast increasing, which is no less than that in model plants of *Arabidopsis thaliana*. Therefore, this review comprehensively and systematically introduced the research progress on melatonin and its function in horticultural crops, and looked forward to the future research to provide some theoretical basis for the application of melatonin in horticultural crop production.

2. The biosynthetic pathway of melatonin in plants

Melatonin is a small molecule, which can shuttle freely in and between cells due to its hydrophilic and lipophilic molecular structure [3]. Using St. John's wort (*Hypericum perforatum* L. cv. Anthos) seedlings as the plant material, Murch et al. [4] exogenously supplied ^{14}C labeled tryptophan and found the presence of radioactive tryptamine, 5-hydroxytryptophan, serotonin, indoleacetic acid and melatonin via adopted isotope tracer approach. Among them, tryptophan and 5-hydroxytryptophan are the synthetic precursors of melatonin in animals. The study indicated that the biosynthetic pathway of melatonin was similar and conserved from animals to plants. Nowadays, the biosynthesis of melatonin using L-tryptophan as substrate has been evidenced in plants with the following steps [5]: L-tryptophan is decarboxylated under the catalysis of L-tryptophan decarboxylase (TDC) to form tryptamine; tryptophan reacts further with tryptophan 5-hydroxylase (T5H) to produce serotonin; serotonin-*N*-acetyltransferase (SNAT) catalyzes the production of *N*-acetylserotonin; and *N*-acetylserotonin can be further catalyzed by *N*-acetylserotonin *O*-methyltransferase (ASMT) or caffeic acid *O*-methyltransferase (CAMT) to produce melatonin (**Figure 1**).

At present, key genes of melatonin synthesis, *TrpDC*, *T5H*, *AcSNMT* and *HIOMT*, have been successfully cloned from plants and their expression patterns have also been carried out [6]. In recent years, plant chloroplasts and mitochondria are considered as the key organelles of the melatonin synthesis [7]. The hypothesis to establish the basis for melatonin is an important biological antioxidant, and chloroplasts and mitochondria are main reactive oxygen species (ROS) sources; these two organelles generate large amounts of melatonin, which is used to remove

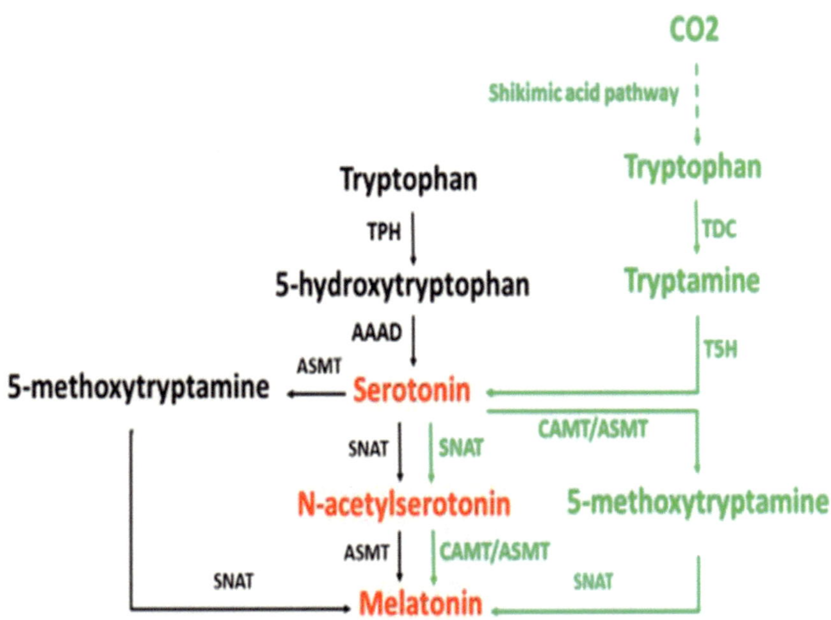

Figure 1.
A comparison of the biosynthetic pathways of melatonin in animals and in plants. Black occurs only in animals; green occurs only in plants; and red occurs in both animals and plants.

excess ROS to reduce the oxidation of cells. However, a recent study in rice (*Oryza sativa* L.) has shown that the key melatonin synthesis gene *COMT* is located in plant chloroplasts and has demonstrated that melatonin synthesis is increased through the 5-MT pathway [8]. CAND2/PMTR1, a hypothetic plant melatonin receptor, recently has been identified in *Arabidopsis thaliana* [9]. It is located in the plasma membrane with a receptor-like topology and interacts with the G-protein subunit (GPA1), while expression in different tissues is induced by melatonin. The binding of plant melatonin receptors triggers Gγb and Gα activation of NADPH oxidase-dependent H_2O_2 production (RBOH), increasing Ca^{2+} inflow, promoting K^+ outflow, and ultimately leading to stomatal closure [10]. The cloning of these melatonin-metabolizing genes and the discovery of the receptors can lay a certain foundation for the study of melatonin-related functions in the future.

3. Concentrations of melatonin in horticultural crops and its influential factors

Although policies regarding the free sale, universal use and food supplement of melatonin are still controversial, most developed countries have classified melatonin as an over-the-counter drug and allowed it to be sold freely in pharmacy stores. Currently, melatonin is regarded as an over-the-counter medicine and health care product to relieve sub-health and improve sleep [11]. Some studies indicated that regular consumption of melatonin-rich foods can significantly improve human health [12–14]. In view of the health benefits of melatonin, more and more nutritionists begin to pay attention to the amount of melatonin in food, hoping that people can get more natural melatonin from daily food.

Two research groups identified melatonin in some edible plants in 1995. Vantassel et al. [15] have identified the presence of plant-derived melatonin in higher plants of morning glory (*Ipomoea nil* L.) and tomato (*Solanum lycopersicum* L.) by radio immunoprecipitation and gas chromatography-mass spectrometry analysis; this was also the first report of plant-derived melatonin in horticultural crops. Since then, melatonin has been isolated and identified in numbers of horticultural crops (**Table 1**). Melatonin levels are usually high in seeds and low in fruits in these edible organs of horticultural crops. According to the statistics, melatonin concentrations in various tissues of plants generally conform to the following rules from high to low: seeds, leaves, roots, flowers and fruits [16]. Moreover, the melatonin concentrations are not evenly distributed in the same tissue. Using white lupine (*Lupinus micranthus* Guss.) leaves as an example, the highest concentrations of melatonin are in the leaf tip, then in the middle leaf, and the least in the leaf base [17]. Through comparative analysis, the distributed gradient of melatonin is similar to the distribution of auxin, which indicates that these two indoleamine compounds may play similar or synergistic roles in plants [18].

The melatonin concentrations in horticultural crops are closely influenced by species, varieties, growing environment, cultivated methods, harvesting time and processing methods. As shown in **Table 1**, melatonin concentration in sweet cherry (13.46 ng g^{-1}) is threefold higher than that in tomato. (4.1 ng g^{-1}) [19]. In different varieties of tomato, the melatonin concentrations fluctuated greatly from 0.5 pg g^{-1} to 114.5 ng g^{-1}. This difference is largely influenced by the genotype of the variety itself. Climate and environmental factors in different years have significant effects on the melatonin concentrations of horticultural crops. For example, the melatonin concentrations of 'Marbone' tomato harvested in 2010 were six times higher than those in 2009, while the melatonin concentrations of 'Festival' strawberries harvested in 2010 were three times lower than those

Common name	Specie name	Melatonin content (pg g^{-1})
Sweet cherries	*Prunus avium* L.	8000–120,000 (FW)
Tart cherries	*Prunus cerasus* L.	1000–19,500 (FW)
White radish	*Raphanus sativus* L.	657.2 (FW)
Ginger	*Zingiber officinale* Rose	583.7 (FW)
Pomegranate	*Punica granatum* L.	540–5500 (FW)
Shungiku	*Chrysanthemum coronarium* L.	416.8 (FW)
Pineapple	*Ananas comosus* L.	302 (FW)
Chinese cabbage	*Brassica rapa* L.	112.5 (FW)
Cabbage	*Brassica oleracea capitata* L.	107.4 (FW)
Carrot	*Daucus carota* L. var. sativa Hoffm.	55.3 (FW)
Taro	*Colocasia esculenta* L.	54.6 (FW)
Apple	*Malus domestica* Borkh.	47.6 (FW)
Spinach	*Basella alba* L.	38.7 (FW)
Onion	*Allium cepa* L.	31.5 (FW)
Cucumber	*Cucumis sativus* L.	24.6 (FW)
Kiwi fruit	*Actinidia Chinensis*	24.4 (FW)
Strawberry	*Fragaria* x *ananassa* Duch.	12.4 (FW)
Asparagus	*Asparagus officinalis* L.	9.5 (FW)
Banana	*Musa acuminata* Colla	8.9 (FW)
Beet root	*Beta vulgaris* L.	2 (FW)
Tomato	*Solanum lycopersicum* L.	0.5–114,500 (FW)
Thyme	*Thymus vulgaris* L.	38,000 (DW)
Chinese liquorice	*Glycyrrhiza uralensis* Fisch.	34,000 (DW)
Coffee beans	*Coffea* sp.	5800–6800 (DW)
Feverfew	*Tanacetum parthenium* L.	1700 (DW)
Mulberry Morus	*Morus alba* L.	1510 (DW)
Black pepper	*Piper nigrum* L.	1092 (DW)
Kidney bean sprouts	*Phaseolus vulgaris* L.	529 (DW)
Aloe	*Aloe vera* L.	516 (DW)
White radish Raphanus	*sativus* L.	485 (DW)
Jujube	*Ziziphus jujube* Lam.	256 (DW)
White mustard seed	Brassica hirta L.	189 (DW)
Qin Jiao	*Gentiana macrophylla* Pall.	180 (DW)
Mustard seed	*Brassica nigra* L.	129 (DW)
Goji berry	*Lycium barbarum* L.	103–530 (DW)
Almond seed	*Prunus amygdalus* Batsch	39 (DW)
Sunflower seed	*Helianthus annuus* L.	29 (DW)
Anise seed	*Pimpinella anisum* L.	7 (DW)
Coriander seed	*Coriandrum sativum* L.	7 (DW)
Celery seed	*Apium graveolens* L.	7 (DW)

Common name	Specie name	Melatonin content (pg g^{-1})
Walnut	*Juglans regia* L.	3.5 (DW)
Bell pepper	*Capsicum annuum* L.	0.179–0.581 (DW)

Table 1.
The content of melatonin in horticultural crops [16].

in 2009 [20]. The concentrations of melatonin in field-cultivated tomato were significantly higher than those in the phytotron-cultivated tomato, and higher than those in the vitro-cultivated tomato [21]. Riga et al. [22] found that fruit bagging can significantly increase the melatonin concentrations in most tomato varieties but reduce the concentrations of melatonin in pepper (*Capsicum annuum* L.). The dynamic change of melatonin concentrations in mammals is regulated by photoperiod and circadian rhythm; the change of melatonin concentrations in plants is also following an analogous pattern. Thus, the harvest time in one day also has an impact on the concentrations of melatonin in the product organs. Zhao et al. [23] showed that the melatonin concentrations of cherries picked at night were significantly higher than those when picked at day times. However, the peak value of melatonin concentrations in cherry fruits usually occurred at 14:00, when the temperature was the highest and the light was intense. This result suggests that melatonin is not only responding to the photoperiod but also involved in a photoprotective mechanism and free radicals scavenging against light damage during photosynthesis. Other studies have shown that the harvest time can also affect the melatonin concentrations in the organs of products. The concentrations of melatonin in cherry usually increase with fruit ripening [24]. The melatonin concentrations of mulberry leaves decreased with the maturity process [25]. The concentrations of melatonin initially increased and then decreased with the chili (*Capsicum annuum* L.) maturation [26]. Similarly with mammals, the highest melatonin concentrations usually exist in young tissues, followed by mature tissues, and then in aging tissues, which indicates that melatonin offers juvenile protection or acts as an aging antagonistic substance. Post-harvest processing has a great influence on the concentrations of melatonin in horticultural products. For example, Kirakosyan et al. [27] compared the melatonin concentrations of frozen, freeze-dried powder, juice and dried fruits of cherry. They found that the cherry juice and cherry dried fruits did not contain melatonin. However, melatonin was detected in frozen cherries and freeze-dried powder, and the melatonin concentrations in freeze cherry were significantly higher than those in freeze-dried fruits. These results indicate that melatonin is not stable and easily degrades upon destruction of the cellular structure during form processing. Yeast is beneficial to produce melatonin during grape wine making [28]. The melatonin concentrations of mulberry tea are only 15% of those in fresh leaves when deep processing of mulberry leaves for green tea and black tea takes place [25]. In addition, melatonin can be detected in most Chinese teas, such as Longjing tea and oolong tea. Therefore, the selection of varieties with high melatonin level, appropriate climatic conditions, reasonable use of bagging cultivation technology and determination of appropriate harvesting time is important to maintain the high melatonin concentrations in horticultural crops. For the fruit and vegetable deep processing products, appropriate processing technology should be considered to maximally maintain the natural melatonin in the products. However, there is still a lack of systematic research on the influence of these factors on melatonin concentrations.

4. Roles of melatonin in regulating growth and development of horticultural crops

4.1 Effects of melatonin on growth and yield formation of horticultural crops

Hernández-Ruiz et al. [18] were the first to propose that melatonin was a hormone-like growth regulator in plants. Because melatonin promotes the hypocotyl growth of albino lupine (*Lupinus micranthus* Guss.) in vitro, its action is related to the concentration gradient in plant tissues. The conclusion was consistent within barley (*Hordeum vulgare* L.), wheat (*Triticum aestivum* L.) and sunflower (*Helianthus annuus* L.) [29]. In terms of chemical structure, the side chain of melatonin has no carboxyl group of IAA, and its growth-promoting activity is about 10–50% of the effect of IAA. Moreover, melatonin does not bind to the IAA receptor, indicating that although melatonin and IAA are co-indoleamine compounds, their signal regulation mechanism is quite different. Since no specific binding receptor of melatonin has been found in plants, melatonin cannot be defined as a plant hormone. Nevertheless, more and more studies are focusing on exploring and identifying the functional and signaling networks that melatonin performs in plants.

Byeon and Back [30] obtained transgenic rice with excessive melatonin accumulation via overexpression of sheep *SNAcT* gene in rice. The growth potential of transgenic rice was significantly higher than that of wild rice and showed the characteristics of delayed flowering and reduced yield. This is consistent with the viewpoint mentioned in Section 2 of this report that "melatonin is a kind of anti-aging substance," indicating that melatonin can inhibit reproductive growth and promote nutritional growth. According to this characteristic, we predicted that through the modification of melatonin-related genes or the exogenous spraying or watering melatonin, the horticultural crops with vegetative organs, as products, could effectively obtain higher economic yield, but the specific effect needs further test to verify.

Other studies have shown that 50 μmol L^{-1} of melatonin solution can significantly promote the growth of soybean seedlings and increase the yield of soybean [31]. Liu et al. [32] found that in the late stage of pear fruit development, spraying of 100 μmol L^{-1} of melatonin solution into the pear tree could promote the accumulation of endogenous melatonin in pear fruit. Melatonin increased the size of pear fruit by increasing the net photosynthetic rate and maximum quantum efficiency of photosystem II. During the ripening period, melatonin increased the concentrations of soluble sugar, especially sucrose and sorbitol, which may be the result of improving the accumulation of starch and promoting the expansion of pear fruit, which had a significant effect on increasing yield. In a study on grapes, Meng et al. [33] found that spraying of 100 mol L^{-1} of melatonin solution on young grape fruits could promote the accumulation of endogenous melatonin in grape fruits and promote the expansion of grape fruits, which had a significant effect on increasing production.

From these effects of melatonin on crop growth and yield, it becomes apparent that the promotion effect of melatonin on crop growth is a common feature. The difference was that the increase of endogenous melatonin concentrations reduced the rice yield, while the exogenous melatonin treatment increased the soybean and grape yield. The effect of melatonin on crop yield seems contradictory; however, we think that first, the soybean experiment adopted melatonin in treatment at seedling stage, but no melatonin was added at the yield formation stage. However, the melatonin level of transgenic rice was high in the whole growth period, so soybean showed an increase in yield, and rice showed a decreased yield. Second, according to the phenotypic difference, melatonin promoted plant growth in soybean more than it did in rice. The full growth of vegetative body provides more abundant photosynthetic products for reproductive growth. Third, the essence of melatonin inhibiting yield formation lies in inhibiting reproductive growth, that is, seed formation. The berry of grape is developed from ovary, which is similar to tomato and other berry crops. The expansion of ovary in the early stage of fruit development needs the stimulation of IAA to form a strong storage for nutrition. The effects of melatonin and IAA are similar, which may be the specific mechanism of melatonin in promoting the expansion of grape fruit. Fourth, Zhao et al. [34] showed that low concentrations of melatonin (10 μmol L^{-1}) could promote the metabolism of sugars, photosynthesis, loading and transportation of sucrose in the maize phloem, thus promoting the growth of maize plants. However, high concentrations of melatonin (1 mmol L^{-1}) could inhibit sucrose loading in phloem, resulting in the accumulation of excessive sucrose, hexose and starch in leaves. As a result, leaf photosynthesis and the growth of maize plants were inhibited. Zhang et al. [35] also obtained similar results on the effects of melatonin concentrations on flowering in apple tree. Overexpression of *SNAcT* gene in rice is similar to the continuous supplementation of melatonin, while in the soybean and grape trials, the short-term supplementation of melatonin resulted in the difference in melatonin supply. Combined with the results of Zhao et al. [34], we analyzed that the concentrations of melatonin in rice was too high, inhibiting the loading of sugars in the phloem, which leads to reduced transport of sugars to the endosperm to form a low-yielding phenotype. The concentration of melatonin in soybean and grape is suitable, which is conducive to the transport of sugar to endosperm and ovary, thus forming a high-yield phenotype. In conclusion, these four factors may be the main reasons behind the differences in the characterization of melatonin on yield. Based on previous studies, we predicted that the right amount of melatonin could increase the yield of root vegetables, leaves, stems, fruits and fruit vegetables as well as the pomology with flesh as edible organs, while the effect on the yield of cauliflower vegetables (such as cauliflower) and pomology with seeds as edible organs (such as walnuts) remains to be verified. However, different crops have different demands on melatonin in different periods. In view of the importance of yield characters in agricultural production, the influence of melatonin on the yield of horticultural crops should become a scientific research hotspot in the future.

4.2 Melatonin regulates the ripening, aging and preservation of horticultural crops

Ripening, aging and preservation are contradictory processes in the production of horticultural products. Interestingly, more and more studies have shown that melatonin can be used to flexibly regulate the ripening, aging and storage of horticultural crops under different concentrations and conditions. Sun et al. [36] found that pretreatment of tomato with 50 μmol L^{-1} and 100 μmol L^{-1} melatonin could promote tomato ripening. Its regulatory mechanism can be summarized as follows: melatonin activates the expression of *ACS4* gene and enhances the synthesis of ethylene. Ethylene signal promoted the synthesis of lycopene by expression of *PSY1* gene through the signal transduction pathways such as *NR*, *ETR4*, *EILs* and *ERF2*, which made the tomato turn color, regulated the expression of *TBG4*, *PG2A*, *Exp1*, *XTH5* and *PE1* genes to degrade the cell wall, softened the fruit, regulated the expression of *PIPs* gene to control fruit water loss, and regulated the expression of *ADH2* and *AAT* genes to promote sugar conversion. Thus, phenotypic characteristics promoting tomato ripening were formed. Similar results have been found in grapes [37]. Lei et al. [38] showed that the synthesis of melatonin in the ripening process of apple fruits was mainly catalyzed by L-tryptophan decarboxylase (TrpDC), tryptophan hydroxylase (T5H), 5-hydroxytryptophan *N*-acetyltransferase (SNAcT) and *N*-acetyl-5-hydroxytryptophan methyltransferase (AcSNMT). Moreover, the research group proposed that during the ripening process of apple fruits, melatonin and malondialdehyde (MDA) concentrations were always negatively correlated. So they speculated that the main role of melatonin in the ripening process of apple fruits was ROS scavenging. When different concentrations of melatonin were used to treat fruits of different species and ripeness, the results were often opposite. Liu et al. [39] found that treatment with melatonin after harvest effectively delayed the senescence of strawberry fruits and reduced the rate of decay, severity of decay and weight loss of strawberry fruits. The main mechanism is to reduce the total soluble solid, H_2O_2 and MDA concentrations of the fruit, promote the accumulation of total phenolic substances and flavonoids, improve the antioxidant capacity of the fruit, delay its color development, and maintain the hardness and titratable acidity.

Shi et al. [40] showed that the concentrations of endogenous melatonin *Arabidopsis thaliana* increased continuously during the process of seedling, maturation and aging. Exogenous spray of melatonin could significantly inhibit the aging process of *Arabidopsis thaliana*. In view of the close relationship between melatonin and IAA, they found that melatonin reduced the expression of *AXR3* and *IAA17* genes, antagonized by IAA signal, which induced the expression of *SEN4* and *SAG12* genes and led to aging. Thus, high concentration of melatonin inhibits the aging process of plants by reducing the expression of *IAA17* and related genes. Arnao and Hernández-Ruiz [41] found that melatonin treatment (1 mmol L^{-1}) significantly inhibited the aging and chlorophyll degradation of barley leaves. Through hormone simulation experiment, the research group believed that the main mechanism of melatonin in inhibiting leaf senescence was synergistic kinetin (KT) and antagonistic abscisic acid (ABA). Wang et al. [42] showed that long-term irrigation of melatonin significantly inhibited the aging of apple leaves. Through proteomics analysis, it was found that melatonin inhibited the activities of most hydrolytic enzymes in plasmids of apple leaves, which were involved in the hydrolysis, redox and stress response, transcriptional regulation, photosynthesis and other senescence-related processes of macromolecular proteins.

4.3 Effects of melatonin on root development of horticultural crops

Endogenous melatonin has similar physiological functions with IAA, which promotes root development, elongation and lateral and adventitious root development. Chen et al. [43] found that 0.1 μmol L^{-1} melatonin promoted the elongation of mustard (*Brassica juncea* L.) root, while 100 μmol L^{-1} melatonin could inhibit its elongation, which was consistent with the physiological concentrations of IAA. The results of IAA concentrations' determination showed that melatonin could induce the accumulation of IAA in mustard roots, which indicated that there was an interaction between melatonin and IAA. In *Arabidopsis thaliana*, melatonin increases the appearance of adventurous roots twofold and the appearance of lateral roots by up to threefold, but has no effect on root hair density [44]. In addition, a large number of lateral roots were induced in three *Arabidopsis thaliana* transgenic lines that produced excessive melatonin compared with the WT lines [45]. Zhang et al. [46] showed that exogenous application of 500 μmol L^{-1} melatonin significantly promoted the occurrence of lateral roots of cucumber. Moreover, transcriptome analysis showed that melatonin caused the upregulation of 121 genes and downregulation of 196 genes. Through GO and pathway enrichment analysis, Zhang et al. believed that melatonin could promote lateral root development of cucumber by activating root-related hormone and transcription factor pathways and reducing oxidative damage caused by respiration during root genesis. Exogenous melatonin could promote the accumulation of nitric oxide (NO) in the periderm of young stem and the tip of new adventitious root, indicating that melatonin-induced NO may be involved in the regulation of adventitious root regeneration. In addition, exogenous application of 50 μmol L^{-1} melatonin significantly promotes the regeneration of tomato adventitious roots [47]. NO acts as the common downstream signal of melatonin and IAA in the process of melatonin-induced root generation by promoting the synthesis, polar transport and hormone signal perception of IAA.

Although melatonin is closely related to IAA in the process of affecting plant root development, melatonin-induced root morphogenesis is independent of auxin signaling [44]. Melatonin promotes the elongation of the principal root and lateral root in *Arabidopsis thaliana*. However, melatonin and IAA have no signal cross-linking and independent signal transduction pathways during root development via auxin signal response label DR5: *uidA* approach. It can be seen that the interaction between melatonin and IAA, two indolamines with similar chemical structure, is extremely complex. Therefore, the following conclusions were made: melatonin is similar to IAA in the process of affecting plant root construction depending on obvious concentration effect. The melatonin-induced root development is relatively independent of IAA in the signal transduction pathway. However, melatonin can promote root development through IAA synthesis, polar transport and hormone perception. In addition, melatonin plays key role in ROS scavenging for the vigorous development of root system, which ensures a good redox balance and the smooth progress of metabolism in cell. The effects of melatonin on root growth make it a broad application prospect in horticulture industry. For example, cucumber and watermelon root usually develop weakly. Thus, pumpkin or gourd is often used as grafting rootstock in these vegetable productions. Although rootstock application enhances root activity, it tends to decrease the sensory quality of the fruit. If proper amount of melatonin is applied to cucumber and watermelon, it may be a good substitute technology for grafting. The application of melatonin to promote adventitious root production is more extensive, such as the grafting technology of "double broken root" for melon and some other fruit vegetables, the induction of root buds for asexual propagation materials, such as potato (*Solanum tuberosum* L.), the cuttage propagation for fruit trees, vegetables and flowers, and the root induction for tissue culture of horticultural crops.

5. Role of melatonin in stress response of horticultural crops

Large numbers of studies have shown that the endogenous melatonin of plants often changes greatly under the stimulation of stress factors, including light intensity, light quality, temperature, water and oxygen, as well as the stimulation of salinity, ultraviolet (UV-B), paraquat, diseases and insect pests, etc. At the same time, exogenous addition of melatonin or enhancement of plant endogenous melatonin synthesis, through gene editing technology, can improve the plant's resistance to adversity.

5.1 Regulation of melatonin resistance to abiotic stress

The efficient utilization of light energy for plant growth, development, yield and quality by light has always been the core of scientific research in horticulture production. Melatonin, on the other hand, has been shown to have a circadian rhythm in mammals, so it is inferred that there should also be a myriad of links between endogenous melatonin of plant and light environmental factors. Melatonin improved the growth performance of yeast under UV radiation and reduced the mortality [48]. The mechanism proposed suggests that melatonin increased the expression of antioxidant genes and DNA-repairing genes.

Kolar et al. [49] used 100 mmol L^{-1} and 500 mmol L^{-1} of melatonin solutions to treat the cotyledon and germ of the short-day plant *Chenopodium rubrum* (*Chenopodium rubrum* L.), which could significantly inhibit the flowering rate under the short-term conditions, and only applied melatonin before the end of the light treatment or the first half of the dark treatment was effective. This suggests that melatonin levels are influenced by circadian rhythms and regulate the photoperiod of plants, which in turn affects some early steps in flowering. In addition, the application of shading technology in the cultivation of capsicum can significantly reduce the melatonin level of capsicum fruits, which indicates that solar radiation can cause the increase of melatonin concentrations [50]. Studies in wheat also show that the concentrations of melatonin in leaves under light condition are significantly higher than they are under dark condition. At the same time, the light treatment significantly increases the concentrations of melatonin synthesis precursors (tryptamine, 5-hydroxytryptamine and *N*-acetyl-5-hydroxytryptamine) in leaves [36]. This suggests that light promotes plants to convert L-tryptophan into melatonin. Light quality also affects the endogenous melatonin concentrations of licorice (*Glycyrrhiza uralensis* L.), following the rule that red light > blue light > white light [51]. The induced effect of light on melatonin is consistent with the hypothesis that chloroplast is the key organelle for melatonin synthesis [52], which is based on the following: pepper peel and wheat leaves are rich in chloroplasts, while light will inevitably produce excess light energy while promoting plant photosynthesis, resulting in the combination of free electrons and O_2 to form ROS. In addition, single light masses, such as red light and blue light, may specifically increase the conversion efficiency of light energy to electric energy at photosystem I (PSI) or photosystem II (PSII), thus increasing the imbalance of electron transfer between PSII and PSI, creating conditions for free electrons to combine with O_2 to generate ROS. Since melatonin is an important antioxidant in organisms, it is of certain biological significance to induce chloroplasts to produce a large amount of melatonin to remove excess ROS so as to reduce the oxidative pressure faced by cells. Unfortunately, at present, there is no direct evidence for the systematic studies about the light quality and light intensity that affect melatonin concentrations for alleviating the strong light damage in plants. Therefore, the mechanism of melatonin metabolism on light in plants and melatonin response to light needs further exploration and verification.

Temperature is a key environmental factor that affects horticultural crops, especially vegetable crops that are planted off season. Inappropriate temperature leads to substantial loss of yield and poor quality of horticultural crops. For this reason, horticultural researchers have been working on temperature adaptation mechanisms and efficient and safe plant growth regulators to cope with sudden temperature changes. Shi et al. [53] confirmed that melatonin was induced by high temperature in *Arabidopsis*, and the application of 20 μmol L^{-1} melatonin significantly improved the expression of heat shock factor (*HSFA1s* and *HSFA2s*) and heat shock protein (*HSP90* and *HSP101*), which led to increased survival rate of *Arabidopsis* under high temperature stress. Jia et al. found that application of 29.0 mg L^{-1} melatonin can enhance the high temperature stress tolerance of cherry radish (*Raphanus sativus* L. var. *radculus pers*), which increased the biomass to 12.9% and the soluble protein to 18.7%. At the same time, the activity of antioxidant enzyme, especially for POD, was enhanced and the lipid peroxidation was reduced under adverse conditions [54]. For melatonin-regulated plant cold tolerance, Lei et al. [55] found that very low concentration of exogenous melatonin (21.5 nmol L^{-1}) could significantly improve the cell viability of carrots, under low-temperature stress, enhance the stability of cell membrane structure, and inhibit orderly degradation of DNA caused by programmed cell death. When tomato was treated with low temperature, the melatonin concentrations and the expression of synthetic control genes were significantly increased [56]. With the extension of stress time, the melatonin concentrations showed an increasing trend, indicating that melatonin played an important role in plant resistance to low-temperature stress. Appropriate osmotic stress can improve the germination rate of cucumber seeds under low-temperature stress [57]. At the same time, osmotic stress intensity was positively correlated with endogenous melatonin concentrations in cucumber seed germination. Further studies showed that the endogenous melatonin, induced by osmotic stress, was beneficial to remove peroxidation damage and stabilize membrane structure under low-temperature stress. However, the excessive endogenous melatonin induced by hyperosmotic stress destroyed the oxidation equilibrium state of protein, but reduced the resistance of cucumber shoots to low temperature. In tomato, the maximum quantum yield (Fv/Fm) of PSII was significantly reduced during chilling, which was effectively alleviated by exogenous melatonin. This is because melatonin induces the expression of violaxanthin de-epoxidase gene, enhances the enzyme activity of violaxanthin de-epoxidase and the enhancement of de-epoxidation state of xanthophyll pigments, promotes the non-photochemical quenching and alleviates the photoinhibition during chilling [58]. In addition, studies on the improvement of plant cold tolerance by exogenous melatonin have also been reported in Kiwifruit (*Actinidia Chinensis*) and *Arabidopsis thaliana*. Wang et al. [59] found that the treatment of Kiwifruit seedlings with 100 μmol L^{-1} melatonin could significantly relieve the growth inhibition and chlorophyll degradation, improve antioxidant enzyme activity and eliminate ROS under low-temperature stress. Exogenous melatonin can induce the expression of a series of low-temperature responsive transcription factors (*CBFs*, *DREBs*, *COR15a*, *CAMTA1* and *ZATs*) [60], which indicates that melatonin has a physiological function in responding to low temperature and transcriptional activation of related metabolic processes.

Horticultural crops require lots of water in their cultivation. Water stress or physiological drought will affect the growth and development of crops and make a significant impact on the yield. Exogenous MT promoted the accumulation of soluble sugar and protein under stress, thereby alleviating the damage of rapeseed seedlings under drought stress [61]. Many orchards in arid/semi-arid areas (especially in mountainous areas) are in a state of long-term water shortage. Although fruit trees can grow, their productions are affected. Melatonin treatment significantly improved the drought resistance of wheat seedlings, including reduced membrane damage, more complete chloroplast grana lamella, higher photosynthetic rate, maximum efficiency of photosystem II, and higher cellular turgor and water-holding capacity [62]. Zuo et al. [45] cloned *AcSNMT*, a key gene for drought-induced melatonin synthesis, from drought-tolerant apple rootstock (*Malus zumi* Mats) and heterologous expressed it in *Arabidopsis thaliana*. The subcellular localization analysis showed that *AcSNMT* gene was mainly located in the nucleus and cell membrane, and the synthesized melatonin could effectively remove drought-induced ROS and improve the growth potential and survival rate of transgenic plants under drought stress. Gong et al. [63] also verified the regulation of melatonin on ROS metabolism under drought stress and the drought-resistant mechanism in tomato. In addition, Meng et al. [64] also found that melatonin could protect the chloroplast membrane structure and grana lamella structure of grape under drought stress, increase the thickness of leaves and the tightness of palisade tissue, and regulate stomatal closure to reduce water loss.

There are about 831 million hectares of saline-alkali land in the world, including 397 million hectares of neutral saline soil and 434 million hectares of alkaline saline soil, accounting for 10% of the world's arable land [65]. Salinity stress can lead to the reduction of water availability and nutrient imbalance, seriously restricting agricultural production. In addition, facility horticulture is also faced with the problem of soil secondary salinization due to the closed environment, lack of rain water leaching in the soil, excessive fertilization and other factors. Therefore, how to improve the salinity tolerance of horticultural crops becomes a key link in the development of characteristic horticulture industry in saline and alkaline areas. Ke et al. [66] demonstrated that melatonin pretreatment regulated polyamine metabolism in wheat and reduced the damage of salt stress. They also believed that melatonin induces enzyme activity that stimulates ROS to clear antioxidant defenses in response to salinity. In addition, exogenous melatonin can also prevent the accumulation of triacylglycerol and promote fatty acid β-oxidation and energy conversion under salt stress conditions. So it is helpful for improving PM H^+-ATPase activity, activating gene expression of Na^+-K^+ reverse transporter, and maintaining K^+/Na^+ homeostasis of sweet potato (*Solanum tuberosum* L.) [67]. Application of 0.5 μmol L^{-1} exogenous melatonin can alleviate symptoms of green leaf loss, reduce the accumulation of Na^+, improve phenols, ascorbic acid and glutathione etc., and promote the related activity of antioxidant enzymes, to alleviate the oxidative damage in salt-stressed tomato [68]. Similarly, root treatment with melatonin alleviated the damage of photosynthetic capacity and oxidative stress and improved antioxidant enzyme activity in salt-stressed watermelon [69]. Zhang et al. [70] also demonstrated in cucumber that exogenous melatonin can upregulate the gibberellin (GA) signaling pathway through the upregulated GA biosynthesis genes (*GA20ox* and *GA3ox*) and inhibit abscisic acid (ABA) signaling pathway through the upregulation of ABA catabolism genes (*CYP707A1* and *CYP707A2*) and downregulation of an ABA biosynthesis gene (*NECD2*), thus promoting the germination rate of cucumber seeds under salt stress. The accumulation of endogenous melatonin in sunflower can be induced by salt stress for 48 h, and the distribution of melatonin accumulation induced by salt stress in the root vascular bundles and cortex is also regionalized. For example, the concentrations of melatonin in sunflower cotyledons and oil-rich tissues are significantly higher than those in other tissues [71]. Moreover, exogenous melatonin can promote the elongation of hypocotyl and root growth of sunflower seedlings under salt stress, and alleviate the inhibition of salt stress on root development of sunflower seedlings to some extent. In addition, our study showed that exogenous addition of 0.5 μmol L^{-1} melatonin could significantly improve the biomass of tomato seedlings, protect photosynthetic organs, promote the activity of antioxidant system, and balance the Na^+-K^+ metabolism of tomato plants under the stress of alkaline salt ($NaHCO_3$) [68]. It has been confirmed that melatonin regulates the physiological processes, such as Na^+ detoxification, dehydration resistance, high pH buffering and ROS scavenging through DREB1 a and IAA3 pathways [72]. These basic studies on exogenous melatonin improving the saline-alkali tolerance of horticultural crops provide theoretical support for the innovation of horticultural crop cultivation technology in saline-alkali areas.

5.2 The regulation of melatonin resistance to biological stress

Plants are often attacked by fungi, bacteria, viruses and pests during their growth and development. Under biological stress, plants produce endogenous hormone-regulated responses, such as salicylic acid (SA), jasmonic acid (JA), ethylene (Eth) and abscisic acid (ABA). Studies in recent years have shown that melatonin can interact with the signaling pathways of biological stress regulated by SA and JA, and negatively regulate plant resistance to biological stress.

The bacterial disease model strains *Pseudomonas syringe pv.* Tomato DC3000 (DC3000) can induce the accumulation of endogenous melatonin, which may play an important role in plant disease resistance response [73, 74]. When 10 μmol L^{-1} solution of melatonin was sprayed on *Arabidopsis* or tobacco leaves, gene expression of disease-course related proteins and defense-related genes activated by SA and Eth signals were induced to reduce the incidence of DC3000. However, exogenous melatonin has no disease-resistant induction effect on *Arabidopsis thaliana* SA and Eth signal mutants of *npr1*, *ein2* and *mpk6*, which indicates that melatonin has close communication with SA and Eth signals. In addition, exogenous melatonin can promote the translocation of the inhibitor *NPR1* of pathophysiology-related protein (*PR1*) from the cytoplasm to the nucleus [75]. All these evidences indicate that melatonin acts as the upstream signal of SA to activate the disease-resistant response chain in plants.

Application of exogenous melatonin to cotton could induce the expression of phenylpropanoid mevalonate (MVA), gossypol and other pathway-related genes, thus leading to increased lignin and gossypol concentrations in metabolites of this pathway and thus enhancing the resistance of cotton to Verticillium wilt [76]. Liu et al. [77] found that spraying of 50 mol L^{-1} of melatonin solution on tomato fruits can induce and enhance the activity of disease-resistant proteases CHI, GLU, PAL and PPO, and significantly enhance disease resistance to *Botrytis cinerea* by regulating ROS accumulation and JA defense signaling pathways. Shi et al. [78] showed that DC3000 could promote the accumulation of melatonin and NO in *Arabidopsis thaliana*, while exogenous melatonin could induce the production of NO. Moreover, both exogenous melatonin and NO treatment of *Arabidopsis thaliana* can enhance its disease resistance and activate the expression of defense genes related to SA signal. This indicates that NO, as the downstream signal of melatonin, acts as the second messenger, communicates the signal network between melatonin and SA, and activates the disease-resistant regulatory network of plants.

In addition to the disease resistance mechanism mediated by SA signal, there is mechanical disease resistance that consists of epidermal tissue, cell wall, phenylalanine pathway-mediated disease resistance mechanism, etc. in plants. Zhao et al. [79] indicated that exogenous melatonin could downregulate the expression of invertase inhibitors of *Arabidopsis thaliana*, activate cell wall invertase activity, promote sucrose metabolism, and accumulate hexose. However, melatonin-induced cell wall invertase activity can increase the strength of cell wall through synthesis of cellulose, xylose and galactose and promote the deposition of callose in cell wall. Melatonin can improve the disease resistance of plants by improving cell wall composition and structure. In the same year, the group of Yin et al. [80] found that exogenous melatonin can reduce marssonina apple blotch and its main regulation mechanism for melatonin can maintain intracellular redox state after infection, improving the phenylalanine ammonia enzyme, polyphenol oxidase, chitinase and glucanase course related to the activity of defense enzymes.

6. Melatonin regulates the signal transduction network of plant growth and stress resistance

As mentioned above, melatonin is widely involved in the regulation of plant growth, development and resistance. Based on this, we sorted out the signal pathways involved in melatonin and summarized the schematic diagram of the signal transduction network regulated by melatonin (**Figure 2**). The main function of melatonin is to promote the biosynthesis of IAA and cooperate with IAA to promote the elongation and expansion of cells, which is manifested in the induction of root growth, lateral root occurrence, adventitious root occurrence and fruit expansion. Although both of them are indoleamine compounds, melatonin and IAA do not share a set of signal transduction networks [44]. Interestingly, melatonin inhibits the expression of auxin antagonistic transcription factor *IAA17*, which blocks auxin signaling [40]. This is equivalent to melatonin indirectly amplifying the IAA signal. In this regard, melatonin and IAA are not absolutely isolated, but have a signal dialogue. In addition, our study also showed that melatonin can promote the polar transport and perception of IAA in tomato root development [47]. In addition, melatonin can induce GA and inhibit ABA synthesis during seed germination [70]. The physiological effects of GA are mainly growth stimulation and aging delay, while ABA is an aging induction hormone. Unfortunately, the relationship of melatonin-GA-ABA is currently limited to germination experiments, and no more conclusions have been drawn to support its regulatory mechanism. Researchers have recently revealed the antioxidant function of melatonin and the inhibitory function of *IAA17* in plant anti-aging process [40, 41]. We believe that melatonin can delay the degradation of chlorophyll, maintain a good redox balance in leaf tissue, and synergistically promote the functions of aging antagonistic hormones, which are the main regulatory mechanisms to inhibit the aging of plant leaves. Healthy leaves can carry out photosynthesis better, which explains that melatonin promotes plant growth from the perspective of carbon nutrition and improves the physiological mechanism of yield. Another interesting issue is that melatonin can induce the release of Eth during the ripening process of tomato fruits, which promotes the ripening of fruits and improves the quality of commodities [36]. However, Eth is also a plant senescence-inducing hormone, which seems to contradict the above-mentioned idea. We speculate that melatonin may have spatio-temporal specificity of developmental period and tissues due to its multipathway properties in synthesis, which needs systematic analysis and experimental verification from the composition of promoter elements. Unfortunately, this area of research is still blank.

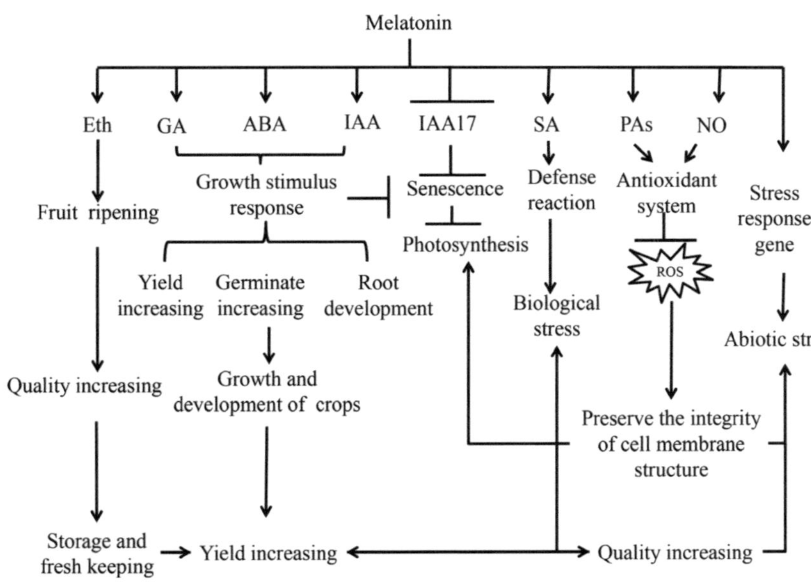

Figure 2.
The signal transduction network of melatonin in plant growth, development and stress tolerance.

Environmental stress and hormones can induce plant cells to produce polyamine (PAs) and NO. However, both PAs and NO are free radicals with strong reactivity. As two active small molecule signaling substances, they are easy to gain and lose electrons. Lei et al. [55] found that melatonin can induce the synthesis of polyamine, under low-temperature stress, and enhance cold resistance in carrot. However, NO is produced by melatonin during the induction of disease resistance [78], alkali resistance [71] and rooting [47], and NO is needed as the downstream signal. In recent years, we have found that NO is the downstream signal of PAs in tomato stress response, and it can activate several plant stress tolerance signaling pathways including antioxidant system [81]. Of course, the mechanical strengthening of melatonin on cell wall tissues [79] and the contact reaction between melatonin and ROS [82] also contributed to the acquisition of plant resilience traits. In addition, melatonin can directly regulate the expression of stress-related functional genes through transcription factor activation, such as senescence-associated genes (SAGs) [41], C-repeat binding factor (*CBFs*) and low-temperature response gene (*COR15a*) [60], heat shock factor (*HSFAs*) and heat shock proteins (*HSPs*) [53], drought response binding factor (*DREBs*) and drought stress resistance genes (*CAMTA1*) [60], plasma membrane Na^+/H^+ reverse transporters (*SOS*), vacuole membrane Na^+/H^+ reverse transporters (*NHX*) and also Na^+ transporters (*HKT*), etc. [59]. In conclusion, these signal transduction network pathways of melatonin systematically enhance the plant's resistance to adversity and improve the survival rate and growth potential of plants under adverse conditions.

7. Conclusion and prospect

Melatonin is widely found in plant tissues, but its concentration in plants is still very low and has obvious tissue specificity. The pan-frying, deep-frying, stir-frying, steaming and stewing techniques commonly used in Chinese food culture are not conducive to the preservation of melatonin in food. Therefore, the food sources mainly focusing on the acquisition of melatonin nutrition are mainly focused on gardening crops like fruits and vegetables suitable for fresh eating. However, melatonin is a substance with similar hormone activity in plants. The use of gene editing technology to comprehensively increase the melatonin concentrations in plants may destroy the balance of melatonin metabolism in plants and bring some adverse effects on the growth and development of plants. Based on this, we propose two suggestions for improving the concentrations of melatonin in horticultural crops: (1) inducing the expression of edible organs or specific developmental periods in horticultural crops by using tissue-specific or inducible promoters combined with melatonin synthesis of key genes. For example, tomato E8 promoter could be used to specifically express the key gene of melatonin synthesis in the fruit, and the effect of excessive melatonin accumulation on the plant was reduced on the basis of increasing the concentrations of melatonin in tomato fruit. Another example is that the chemical-induced expression system TetR combined with melatonin synthesis of key genes can be used to induce the expression of horticultural products in a time period prior to harvesting, which can improve the melatonin concentrations of the harvested products. (2) The *Arabidopsis thaliana* mutant library was used to dig out the mutant with excessive melatonin accumulation, locate the key gene and use CRISPR-Cas9 to conduct targeted gene editing on horticultural crops to create a new horticultural germ plasm rich in melatonin.

The study of melatonin can promote the improvement of global ecological environment. In the horticultural production system under great pressure in the natural environment and frequent outbreak of biological stress, the study on melatonin promoting growth and anti-stress regulation mechanism and the successful cases of exogenous melatonin promoting crop growth and enhancing resistance are expected to make great contributions to the high yield and high-quality production of horticultural crops. Based on the current research status, we make a prediction on the application of melatonin in horticulture production, hoping to attract more and more attention. (1) For cucumbers and others prone to premature aging, horticultural crops can be applied by exogenous melatonin to alleviate the aging process. (2) Leek, asparagus and most fruit trees show aging trend with the extension of cultivation year. The application of exogenous melatonin also plays a role in the rejuvenation of this kind of aging horticultural crops. (3) Taking advantage of the anti-aging and drought resistance properties of melatonin, it can be used for fresh cut flowers, fruits and leafy vegetables. (4) Melatonin can promote root growth and withstand low temperature in plants, alleviating the reduction in root growth and absorption of fertilizer and water that are caused by the excessive low ground temperature, thereby reducing heating costs, improving the utilization rate of CO_2, and promoting the efficacy of fertilizer. (5) At present, shade means are often used in the over-summer cultivation of ginger and other light-tolerant and shade-tolerant horticultural crops to reduce leaf senescence and root loss caused by the damage of strong light to leaf photosynthetic organs. Melatonin is introduced into the cultivation technology of this type of crops by utilizing the mechanism of accumulation of melatonin in chloroplast and protection of photosynthetic organs and inhibition of leaf senescence. (6) In cucumber and pumpkin grafting, the technology of double broken root grafting is often adopted. Melatonin can promote the occurrence of adventitious root and prevent the infection of wound. At the same time, melatonin also has the function of promoting graft wound healing, which has great application potential in some fruit trees and flowers with low graft survival rate. (7) Taking advantage of the characteristics of melatonin in delaying senescence, MT is used as a flower-/fruit-preserving agent for vegetable crops in winter and summer to improve pollen vitality, increase fruit setting rate and reduce malformation fruit rate. (8) As a plant growth regulator, melatonin can be used in plant growth promotion and disease prevention to reduce the use of pesticides. Moreover, as a new type of growth regulation substance, it has the characteristics of high efficiency and environmental protection. Its development and utilization can greatly promote the healthy growth of horticultural crops, improve the fertilizer utilization ratio and reduce the occurrence of plant diseases and insect pests and pesticide dosage. It can play an important role in the process of using fewer chemical fertilizers and pesticides in China, and has a great development potential in promoting the healthy development of horticulture industry.

Acknowledgements

The authors would like to acknowledge the National Natural Science Foundation of China [U1903105 and 31872943] and the Natural Science Foundation of Shandong Province [ZR2019MC067] for their support.

Author details

Yanyan Yan[1,2,3,4], Qinghua Shi[1,2,3,4] and Biao Gong[1,2,3,4*]

1 State Key Laboratory of Crop Biology, P.R. China

2 Collaborative Innovation Center of Fruit and Vegetable Quality and Efficient Production in Shandong, P.R. China

3 Key Laboratory of Biology and Genetic Improvement of Horticultural Crops in Huanghuai Region, Ministry of Agriculture and Rural Affairs, P.R. China

4 College of Horticulture Science and Engineering, Shandong Agricultural University, Taian, P.R. China

*Address all correspondence to: gongbiao@sdau.edu.cn

References

[1] Lerner AB, Case JD, Takahashi Y, Lee TH, Mori W. Isolation of melatonin, the pineal gland factor that lightens melanocytes. Journal of the American Chemical Society. 1958;**80**:2587-2587

[2] Balzer I, Hardeland R. Photoperiodism and effects of indoleamines in a unicellular alga, *Gonyaulax polyedra*. Science. 1991;**253**:795-797

[3] Arano MB, Hernández-Ruiz J. Melatonin in plants: More studities are necessary. Plant Signaling and Behavior. 2007;**2**:381-382

[4] Murch SJ, KrishnaRaj S, Saxena PK. Tryptophan is a precursor for melatonin and serotonin biosynthesis in in vitro regenerated St. John's wort (*Hypericum perforatum* L. cv. Anthos) plants. Plant Cell Reports. 2007;**19**:698-704

[5] Tan DX, Manchester LC, Esteban-Zubero E, Zhou Z, Reiter RJ. Melatonin as a potent and inducible endogenous antioxidant: Synthesis and metabolism. Molecules. 2015;**20**:18886-18906

[6] Park WJ. Melatonin as an endogenous plant regulatory signal: Debates and perspectives. Journal of Pineal Research. 2011;**54**:143-149

[7] Tan DX, Manchester LC, Liu X, Rosales-Corral SA, Castroviejo DA, Reiter RJ. Mitochondria and chloroplasts as the original sites of melatonin synthesis: A hypothesis related to melatonin's primary function and evolution in eukaryotes. Journal of Pineal Research. 2013;**54**:127-138

[8] Geun-Hee C, Lee HY, Back K. Chloroplast overexpression of rice caffeic acid O-methyltransferase increases melatonin production in chloroplasts via the 5-methoxytryptamine pathway in transgenic rice plants. Journal of Pineal Research. 2017;**63**:e12412

[9] Wei J, Li DX, Zhang JR, Shan C, Rengel Z, Song ZB, et al. Phytomelatonin receptor PMTR1-mediated signaling regulates stomatal closure in *Arabidopsis thaliana*. Journal of Pineal Research. 2018;**65**:e12500

[10] Arnao MB, Hernández-Ruiz J. Melatonin: A new plant hormone and/or a plant master regulator? Trends in Plant Science. 2019;**24**:38-48

[11] Arnao MB, Hernández-Ruiz J. Phytomelatonin, natural melatonin from plants as a novel dietary supplement: Sources, activities and world market. Journal of Functional Foods. 2018;**48**:37-42

[12] Aridas JDS, Yawno T, Sutherland AE, Nitsos I, Ditchfield M, Wong FY, et al. Systemic and transdermal melatonin administration prevents neuropathology in response to perinatal asphyxia in newborn lambs. Journal of Pineal Research. 2018;**64**:e12479

[13] Fernández VG, Reiter RJ, Agil A. Melatonin increases brown adipose tissue mass and function in Zücker diabetic fatty rats: Implications for obesity control. Journal of Pineal Research. 2018;**64**:e12472

[14] Montalbano G, Mania M, Abbate F, Navarra M, Guerrera MC, Laura R, et al. Melatonin treatment suppresses appetite genes and improves adipose tissue plasticity in diet-induced obese zebrafish. Endocrine. 2018;**62**:381-393

[15] Vantassel DL, Li JA, Oneill SD. Melatonin-identification of a potential dark signal in plants. Plant Physiology. 1993;**102**:117-117

[16] Arnao MB. Phytomelatonin: Discovery, content, and role in plants. Advances in Botany. 2014;**2014**:e815769

[17] Hernández-Ruiz J, Arnao MB. Distribution of melatonin in different zones of lupin and barley plants at different ages in the presence and absence of light. Journal of Agricultural and Food Chemistry. 2008;**56**:10567-10573

[18] Hernández-Ruiz J, Cano A, Arnao MB. Melatonin: A growth-stimulating compound present in lupin tissues. Planta. 2004;**220**:140-144

[19] Burkhardt S, Tan DX, Manchester LC, Hardeland R, Reiter RJ. Detection and quantification of the antioxidant melatonin in Montmorency and Balaton tart cherries (*Prunus cerasus*). Journal of Agricultural and Food Chemistry. 2001;**49**:4898-4902

[20] Stürtz M, Cerezo AB, Cantos-Villar E, Garcia-Parrilla MC. Determination of the melatonin content of different varieties of tomatoes (*Lycopersicon esculentum*) and strawberries (*Fragaria ananassa*). Food Chemistry. 2011;**127**:1329-1334

[21] Arano MB, Hernández-Ruiz J. Growth conditions influence the melatonin content of tomato plants. Food Chemistry. 2013;**138**:1212-1214

[22] Riga P, Medina S, García-Flores LA, Gil-Izquierdo Á. Melatonin content of pepper and tomato fruits: Effects of cultivar and solar radiation. Food Chemistry. 2014;**156**:347-352

[23] Zhao Y, Tan DX, Lei Q, Chen H, Wang L, Li QT, et al. Melatonin and its potential biological functions in the fruits of sweet cherry. Journal of Pineal Research. 2012;**55**:79-88

[24] Zhao Y, Tan DX, Lei Q, Chen H, Wang L, Li QT, et al. Melatonin and its potential biological functions in the fruits of sweet cherry. Journal of Pineal Research. 2012;**55**:79-88

[25] Pothinuch P, Tongchitpakdee S. Melatonin contents in mulberry (*Morus* spp.) leaves: Effects of sample preparation, cultivar, leaf age and tea processing. Food Chemistry. 2011;**128**:415-419

[26] Korkmaz A, Deger O, Cuci Y. Profiling the melatonin content in organs of the pepper plant during different growth stages. Scientia Horticulturae. 2014;**172**:242-247

[27] Kirakosyan A, Seymour EM, Llanes DEU, Kaufman PB, Bolling SF. Chemical profile and antioxidant capacities of tart cherry products. Food Chemistry. 2009;**115**:20-25

[28] Vigentini I, Gardana C, Fracassetti D, Gabrielli M, Foschino R, Simonetti P, et al. Yeast contribution to melatonin, melatonin isomers and tryptophan ethyl ester during alcoholic fermentation of grape musts. Journal of Pineal Research. 2015;**58**:388-396

[29] Arnao MB, Hernández-Ruiz J. Growth activity, rooting capacity, and tropism: Three auxinic precepts fulfilled by melatonin. Acta Physiologiae Plantarum. 2017;**39**:127

[30] Byeon Y, Back K. An increase in melatonin in transgenic rice causes pleiotropic phenotypes, including enhanced seedling growth, delayed flowering, and low grain yield. Journal of Pineal Research. 2014;**56**:408-414

[31] Wei W, Li QT, Chu YN, Reiter RJ, Yu XM, Zhu DH, et al. Melatonin enhances plant growth and abiotic stress tolerance in soybean plants. Journal of Experimental Botany. 2014;**66**:695-707

[32] Liu JL, Yue RR, Si M, Wu M, Cong L, Zhai R, et al. Effects of exogenous application of melatonin on quality and sugar metabolism in 'Zaosu' pear fruit. Journal of Plant Growth Regulation. 2019;**38**:1161-1169

[33] Meng JF, Xu TF, Song CZ, Yu Y, Hu F, Zhang L, et al. Melatonin treatment of pre-veraison grape berries to increase size and synchronicity of berries and modify wine aroma components. Food Chemistry. 2015;**185**:127-134

[34] Zhao H, Su T, Huo L, Wei H, Jiang Y, Xu L, et al. Unveiling the mechanism of melatonin impacts on maize seedling growth: Sugar metabolism as a case. Journal of Pineal Research. 2015;**59**:255-266

[35] Zhang HX, Wang L, Shi K, Shan DQ, Zhu YP, Wang CY, et al. Apple tree flowering is mediated by low level of melatonin under the regulation of seasonal light signal. Journal of Pineal Research. 2019;**66**:e12551

[36] Sun Q, Zhang N, Wang J, Zhang H, Li D, Shi J, et al. Melatonin promotes ripening and improves quality of tomato fruit during postharvest life. Journal of Experimental Botany. 2014;**66**:657-668

[37] Xu LL, Yue Q, Bian F, Sun H, Zhai H, Yao YX. Melatonin enhances phenolics accumulation partially via ethylene signaling and resulted in high antioxidant capacity in grape berries. Frontiers in Plant Science. 2017;**8**:1426

[38] Lei Q, Wang L, Tan DX, Zhao Y, Zheng XD, Chen H, et al. Identification of genes for melatonin synthetic enzymes in 'Red Fuji' apple (*Malus domestica Borkh.cv.Red*) and their expression and melatonin production during fruit development. Journal of Pineal Research. 2013;**55**:443-451

[39] Liu CH, Zheng HH, Sheng KL, Liu W, Zheng L. Effects of melatonin treatment on the postharvest quality of strawberry fruit. Postharvest Biology and Technology. 2018;**139**:47-55

[40] Shi H, Reiter RJ, Tan DX, Chan Z. *INDOLE-3-ACETIC ACID INDUCIBLE 17* positively modulates natural leaf senescence through melatonin-mediated pathway in Arabidopsis. Journal of Pineal Research. 2015;**58**:26-33

[41] Arano MB, Hernández-Ruiz J. Protective effect of melatonin against chlorophyll degradation during the senescence of barley leaves. Journal of Pineal Research. 2009;**46**:58-63

[42] Wang P, Sun X, Xie Y, Li M, Chen W, Zhang S, et al. Melatonin regulates proteomic changes during leaf senescence in *Malus hupehensis*. Journal of Pineal Research. 2014;**57**:291-307

[43] Chen Q, Qi WB, Reiter RJ, Wei W, Wang BM. Exogenously applied melatonin stimulates root growth and raises endogenous indoleacetic acid in roots of etiolated seedlings of *Brassica juncea*. Journal of Plant Physiology. 2009;**166**:324-328

[44] Pelagio-Flores R, Muñoz-Parra E, Ortiz-Castro R, Lopez-Bucio J. Melatonin regulates Arabidopsis root system architecture likely acting independently of auxin signaling. Journal of Pineal Research. 2012;**53**:279-288

[45] Zuo B, Zheng X, He P, Wang L, Lei Q, Feng C, et al. Overexpression of *MzASMT* improves melatonin production and enhances drought tolerance in transgenic *Arabidopsis thaliana* plants. Journal of Pineal Research. 2014;**57**:408-417

[46] Zhang N, Zhang HJ, Zhao B, Sun QQ, Cao YY, Li R, et al. The RNA-seq approach to discriminate gene expression profiles in response to melatonin on cucumber lateral root formation. Journal of Pineal Research. 2014;**56**:39-50

[47] Wen D, Gong B, Sun SS, Liu S, Wang X. Wei M, et al, Promoting roles of melatonin in adventitious root development of *Solanum lycopersicum* L. by regulating auxin and nitric oxide

signaling. Frontiers in Plant Science. 2016;**7**:718

[48] Bisquert R, Muñiz-Calvo S, Guillamón JM. Protective role of intracellular melatonin against oxidative stress and UV radiation in *Saccharomyces cerevisiae*. Frontiers in Microbiology. 2018;**9**:318

[49] Kolář J, Johnson HJ, Macháčková I. Exogenously applied melatonin (*N*-acetyl-5-methoxytryptamine) affects flowering of the short-day plant *Chenopodium rubrum*. Physiology Plantarum. 2003;**118**:605-612

[50] Riga P, Medina S, García-Flores LA, Gil-Izquierdo Á. Melatonin content of pepper and tomato fruits: Effects of cultivar and solar radiation. Food Chemistry. 2014;**156**:347-352

[51] Afreen F, Zobayed SMA, Kozai T. Melatonin in *Glycyrrhiza uralensis*: Response of plant roots to spectral quality of light and UV-B radiation. Journal of Pineal Research. 2006;**41**:108-115

[52] Tan DX, Manchester LC, Liu X, Rosales-Corral SA, Castroviejo DA, Reiter RJ. Mitochondria and chloroplasts as the original sites of melatonin synthesis: A hypothesis related to melatonin's primary function and evolution in eukaryotes. Journal of Pineal Research. 2013;**54**:127-138

[53] Shi H, Tan DX, Reiter RJ, Ye T, Yang F, Chan Z. Melatonin induces class A1 heat-shock factors (HSFA1s) and their possible involvement of thermotolerance in *Arabidopsis*. Journal of Pineal Research. 2015;**58**:335-342

[54] Jia C, Yu X, Zhang M, Liu Z, Zou P, Ma J, et al. Application of melatonin-enhanced tolerance to high-temperature stress in Cherry Radish (*Raphanus sativus L. var. radculus pers*). Journal of Plant Growth Regulation. 2019:1-10

[55] Lei XY, Zhu RY, Zhang GY, Dai YR. Attenuation of cold-induced apoptosis by exogenous melatonin in carrot suspension cells: The possible involvement of polyamines. Journal of Pineal Research. 2004;**36**:126-131

[56] Sharafi Y, Aghdam MS, Luo Z, Jannatizadeh A, Razavi F, Fard JR, et al. Melatonin treatment promotes endogenous melatonin accumulation and triggers GABA shunt pathway activity in tomato fruits during cold storage. Scientia Horticulturae. 2019;**254**:222-227

[57] Posmyk MM, Balabusta M, Wieczorek M, Sliwinska E, Janas KM. Melatonin applied to cucumber (*Cucumis sativus* L.) seeds improves germination during chilling stress. Journal of Pineal Research. 2009;**46**:214-223

[58] Ding F, Wang M, Liu B, Zhang S. Exogenous melatonin mitigates photoinhibition by accelerating non-photochemical quenching in tomato seedlings exposed to moderate light during chilling. Frontiers in Plant Science. 2017;**8**:244

[59] Gao F, Xia H, Yuan XZ, Huang SY, Liu J, Liang D. Effects of exogenous melatonin on phenolic substance content and antioxidant ability of kiwifruit seedlings under salt stress. Acta Agriculturae Zhejiangensis. 2017;**29**:1144-1150

[60] Bajwa VS, Shukla MR, Sherif SM, Murch SJ, Saxena PK. Role of melatonin in alleviating cold stress in Arabidopsis thaliana. Journal of Pineal Research. 2014;**56**:238-245

[61] Li J, Zeng L, Cheng Y, Lu G, Fu G, Ma H, et al. Exogenous melatonin alleviates damage from drought stress in *Brassica napus* L. (rapeseed) seedlings. Acta Physiologiae Plantarum. 2018;**40**:43

[62] Cui G, Zhao X, Liu S, Sun F, Zhang C, Xi Y. Beneficial effects of melatonin in overcoming drought stress in wheat seedlings. Plant Physiology and Biochemistry. 2017;**118**:138-149

[63] Gong B, Yan YY, Wen D, Shi QH. Hydrogen peroxide produced by NADPH oxidase: A novel downstream signaling pathway in melatonin-induced stress toelrance in *Solanum lycopersicum*. Physiologia Plantarum. 2017;**160**:396-409

[64] Meng JF, Xu TF, Wang ZZ, Fang YL, Xi ZM, Zhang ZW. The ameliorative effects of exogenous melatonin on grape cuttings under water-deficient stress: Antioxidant metabolites, leaf anatomy, and chloroplast morphology. Journal of Pineal Research. 2014;**57**:200-212

[65] Gong B, Li X, Vandenlangenberg KM, Wen D, Sun S, Wei M, et al. Overexpression of S-adenosyl-L-methionine synthetase increased tomato tolerance to alkali stress through polyamine metabolism. Plant Biotechnology Journal. 2014;**12**:694-708

[66] Ke Q, Ye J, Wang B, Ren J, Yin L, Deng X, et al. Melatonin mitigates salt stress in wheat seedlings by modulating polyamine metabolism. Frontiers in Plant Science. 2018;**9**:1-11

[67] Yu Y, Wang A, Li X, Kou M, Wang W, Chen X, et al. Melatonin-stimulated triacylglycerol breakdown and energy turnover under salinity stress contributes to the maintenance of plasma membrane H^+-ATPase activity and K^+/Na^+ homeostasis in sweet potato. Frontiers in Plant Science. 2018;**9**:256

[68] Liu N, Gong B, Jin ZY, Wang XF, Wei M, Yang FJ, et al. Sodic alkaline stress mitigation by exogenous melatonin in tomato needs nitric oxide as a downstream signal. Journal of Plant Physiology. 2015;**186-187**:68-77

[69] Li H, Chang J, Chen H, Wang Z, Gu X, Wei C, et al. Exogenous melatonin confers salt stress tolerance to watermelon by improving photosynthesis and redox homeostasis. Frontiers in Plant Science. 2018;**8**:295

[70] Zhang HJ, Zhang N, Yang RC, Wang L, Sun QQ, Li DB, et al. Melatonin promotes seed germination under high salinity by regulating antioxidant systems, ABA and GA4 interaction in cucumber (*Cucumis sativus* L.). Journal of Pineal Research. 2014;**57**:269-279

[71] Mukherjee S, David A, Yadav S, Baluška F, Bhatla SC. Salt stress-induced seedling growth inhibition coincides with differential distribution of serotonin and melatonin in sunflower seedling roots and cotyledons. Physiologia Plantarum. 2014;**152**:714-728

[72] Yan YY, Jing X, Tang HM, Li XT, Gong B, Shi QH. Using transcriptome to discover a novel melatonin-induced sodic alkaline stress reistant pathway in *Solanum lycopersicum* L. Plant Cell Physiology. 2019;**60**:2051-2064

[73] Liu N, Jin Z, Wang S, Gong B, Wen D, Wang X, et al. Sodic alkaline stress mitigation with exogenous melatonin involves reactive oxygen metabolism and ion homeostasis in tomato. Scientia Horticulturae. 2015;**181**:18-25

[74] Lee HY, Byeon Y, Back K. Melatonin as a signal molecule triggering defense responses against pathogen attack in Arabidopsis and tobacco. Journal of Pineal Research. 2014;**57**:262-268

[75] Lee HY, Byeon Y, Tan DX, Reiter RJ, Back K. Arabidopsis serotonin N-acetyltransferase knockout mutant plants exhibit decreased melatonin and salicylic acid levels resulting in susceptibility to an avirulent

pathogen. Journal of Pineal Research. 2015;**58**:291-299

[76] Li C, He QL, Zhang F, Yu JW, Li C, Zhao TL, et al. Melatonin enhances cotton immunity to *Verticillium* wilt via manipulating lignin and gossypol biosynthesis. The Plant Journal. 2019;**100**:784-800

[77] Liu CX, Chen LL, Zhao RR, Li R, Zhang SJ, Yu WQ, et al. Melatonin induces disease resistance to *Botrytis cinerea* in tomato fruit by activating jasmonic acid signaling pathway. Journal of Agricultural and Food Chemistry. 2019;**67**:6116-6124

[78] Shi H, Chen Y, Tan DX, Reiter RJ, Chan Z, He C. Melatonin induces nitric oxide and the potential mechanisms relate to innate immunity against bacterial pathogen infection in Arabidopsis. Journal of Pineal Research. 2015;**59**:102-108

[79] Zhao H, Xu L, Su T, Jiang Y, Hu L, Ma F. Melatonin regulates carbohydrate metabolism and defenses against *Pseudomonas syringae* pv. *tomato DC3000* infection in *Arabidopsis thaliana*. Journal of Pineal Research. 2015;**59**:109-119

[80] Yin L, Wang P, Li M, Ke X, Li C, Liang D, et al. Exogenous melatonin improves *Malus* resistance to Marssonina apple blotch. Journal of Pineal Research. 2013;**54**:426-434

[81] Gong B, Li X, Bloszies S, Wen D, Sun S, Wei M, et al. Sodic alkaline stress mitigation by interaction of nitric oxide and polyamines involves antioxidants and physiological strategies in *Solanum lycopersicum*. Free Radical Biology and Medicine. 2014;**71**:36-48

[82] Zhang N, Sun Q, Zhang H, Cao Y, Weeda S, Ren S, et al. Roles of melatonin in abiotic stress resistance in plants. Journal of Experimental Botany. 2015;**66**:647-656

Section 6

Effect of Melatonin on 5-HT Active Transport

Chapter 8

Studies on Tryptophan Metabolites in Patients of Major Monopolar Depression

Hiroi Tomioka, Junichi Masuda, Akikazu Takada and Akira Iwanami

Abstract

Plasma levels of tryptophan metabolites were compared between healthy volunteers and patients of major monopolar depression at various ages and genders. An ultrahigh-speed liquid chromatography/mass spectrometry has been used for analysis. There are significant gender and age differences in TRP metabolites of healthy volunteers. At the upper stream of metabolism, metabolites of young women and old men are higher, but at the lower stream of metabolism, their levels are higher in young men and old women. Such differences disappear in plasma of patients of major monopolar depression except for kynurenine (KYN). Daily variation of blood serotonin (5-HT) levels showed that 5-HT levels were low in the morning and increased toward evening, but blood levels of 5-HT were higher in healthy people than depressive people in the morning and decreased to ward evening. Significant age and gender differences of plasma levels of tryptophan metabolites in healthy volunteers disappear in patients of major monopolar depression. Blood levels of 5-HT were higher in healthy people than depressive patients.

Keywords: depression, monopolar depression, bipolar depression, tryptophan, serotonin, 5-hydroxyindoleacetic acid, kynurenine, 3-hydroxykynurenine, kynurenic acid, anthranilic acid, xanthurenic acid, indole-3-acetic acid, selective serotonin reuptake inhibitor (SSRI), serotonin norepinephrine reuptake inhibitor (SNRI), anxiolytic, antipsychotic, and circadian variation

1. Introduction

Recently, it has been shown that both the responses to placebo and antidepressant increased [1]. Kirsch has claimed that pharmaceutical companies did not include mildly and moderately depressed patients in trials of efficacy after finding that these patients did not benefit beyond placebo [2]. He consistently insists that antidepressants are not more effective than placebos in moderately depressed patients [3]. Drug-placebo differences are considered to be small in efficacy trials, and most of the response to antidepressants seems due to expectancy [3].

Major depressive disorder is one of the most common psychiatric disorders which is burdensome and costly worldwide in adults. Although pharmacological and non-pharmacological treatments are available, because of inadequate resources, antidepressants are used more frequently than psychological interventions.

By using a meta-analysis, all antidepressants were shown more efficacious than placebo in adults with major depressive disorder. Smaller differences between active drugs were found when placebo-controlled trials were included in the analysis [4].

Serotonin (5-HT) has been indicated to be involved in etiology of depression [5].

The roles of various metabolites of kynurenine (KYN) pathway are reviewed [6], so we do not discuss these roles in detail.

As to relationships between serotonin levels and depression, we analyzed plasma levels of TRP metabolites in patients of depression.

Although the concentration of 5-HT has been considered to be low in depressive patients [7], 5-HT concentration in the brains of suicide victims were not low [8]. Therefore, it is not known if 5-HT concentration is decreased in the brains of depressive patients.

We have recently succeeded in simultaneous measurements of TRP metabolites in plasma using an ultrahigh-speed liquid chromatography/mass spectrometry (LC/MS) [9–12].

We now report age and gender differences of various TRP metabolites in patients of major monopolar depression and healthy volunteers.

2. Results

2.1 Comparison of plasma levels of TRP metabolites between healthy people and patients of major depression

2.1.1 Healthy volunteers

The sample sizes and ages of participants are as follows. Old men (n = 25; age, 60.8 ± 9.9) and old women (n = 39; age, 67.4 ± 7.5) and young men (n = 49; age, 20.7 ± 1.5) and young women (n = 47; age, 21.2 ± 0.7). Characteristics of these people are described in **Table 1**.

Subjects	young men n=48 a	young women n=47 b	old men n=44 c	old women n=39 d
Age (years)	20.7±1.5	21.2±0.7	62.4±9.6	67.4±7.5
Height (m)	1.72±0.06	1.58±0.06	1.68±0.07	1.57±0.06
Weight (kg)	65.1±9.2	50.9±5.8	68.8±10.9	50.6±6.8
BMI	22.1±3.2	20.3±1.6	24.3±3.2	20.5±2.5

Table 1.
Background of healthy participants.

2.1.2 Patients

Outpatients of depression were recruited in this study. Fasting blood samples were taken early in the morning. Their severity of depression was checked by clinical global impression—severity scale (CGI-S), SRS, and Hamilton depression rating scale (HDRS). The history of prescriptions of drugs such as antidepressants, anxiolytics, mood stabilizers, and other drugs were asked.

Sample numbers are 55 (male, 15; female, 40; average age, 45.4 ± 11.9). The number of MDD is 38 and BD is 17. Further characteristics of patients are described below.

Plasma factors were measured after plasma was separated from blood (3000 rpm/min at 4°C). Ethylenediaminetetraacetic acid (EDTA) was used as an anticoagulant.

2.1.3 The simultaneous measurements of TRP metabolites in plasma

An ultrahigh-speed liquid chromatography/spectrometry was used for the assay. Although detailed methodology was described elsewhere [5–9], the important improvement of the assay method is described here.

2.1.3.1 Reagents and instrumentation

The simultaneous analytical method developed can be adapted to major metabolites of TRP including melatonin in clinical sample.

The analytical targets of developed method are major metabolites, such as tryptophan (TRP), L-5-hydroxytryptophan (5-HTP), serotonin (5-HT), kynurenine (KYN), 5-hydroxy-tryptophol, tryptophol, 5-hydroxyindoleacetic acid (5-HIAA), indole-3-acetic acid, anthranilic acid (AA), kynurenic acid (KYNA), quinaldic acid, indole-3-butyric acid, 3-hydroxykynurenine (3-HKYN), 3-hydroxyanthranilic acid (3-HAA), xanthurenic acid (XA), melatonin, and quinolinic acid (QA). Each compound was purchased from major chemical regent manufacturers, such as FUJIFILM Wako chemical (Osaka, Japan) and Sigma-Aldrich (St. Louis, MO, USA).

Metabolite analysis was performed by a liquid chromatograph tandem mass spectrometer, the LCMS-8060 quadrupole mass spectrometer combined with Nexera X2 liquid chromatograph system (Shimadzu Corporation, Kyoto, Japan).

The targets are separated by reversed-phase chromatography using C18 analytical column, L-Columns ODS2 (2.1 mm × 150 mm, CERI, Tokyo, Japan) with a gradient elution. Mobile phases were 0.1% formic acid solution and acetonitrile with the gradient elution by 5% concentration of acetonitrile in 3 min and then 5–95% in 6 min, followed by 5% in 3 min at a total flow rate of 0.4 mL/min. The temperature of the column was 40°C. Electrospray ionization (ESI) was used as mostly positive ionization with multi-reaction monitoring (MRM) detection.

Flow rate of the neutralizer and the drying gas were 2 L/min and 10 mL/min, respectively. Temperature of desolvation line (heated capitally tube) was 250°C. ESI interface was used at 400°C with 10 L/min of heating gas flow. Each MRM transition was optimized using each standard solution. Optimized results were shown in **Table 1**.

All mother solutions of 1 mg/mL had been stocked under −80°C, and standard samples for calibration curve were prepared prior to use as mixture solution by consideration of each range of measurement concentration.

2.1.3.2 Analysis of human plasma

Aliquot of 50 μL human plasma was used for each sample analysis. The procedure including deproteinization is shown in **Figure 1**.

TRP metabolic pathways are shown in **Figure 1**.

Figure 1 shows metabolic pathways of TRP. Metabolites were measured by an ultrahigh-speed liquid chromatography/mass spectrometry.

One-way ANOVA was used for evaluating statistical significance. A, b, c, and d indicate values of young and old men and women. Tukey's test was used for post hoc test.

Table 2 shows that there are significant gender and age differences in plasma levels of TRP of healthy volunteers.

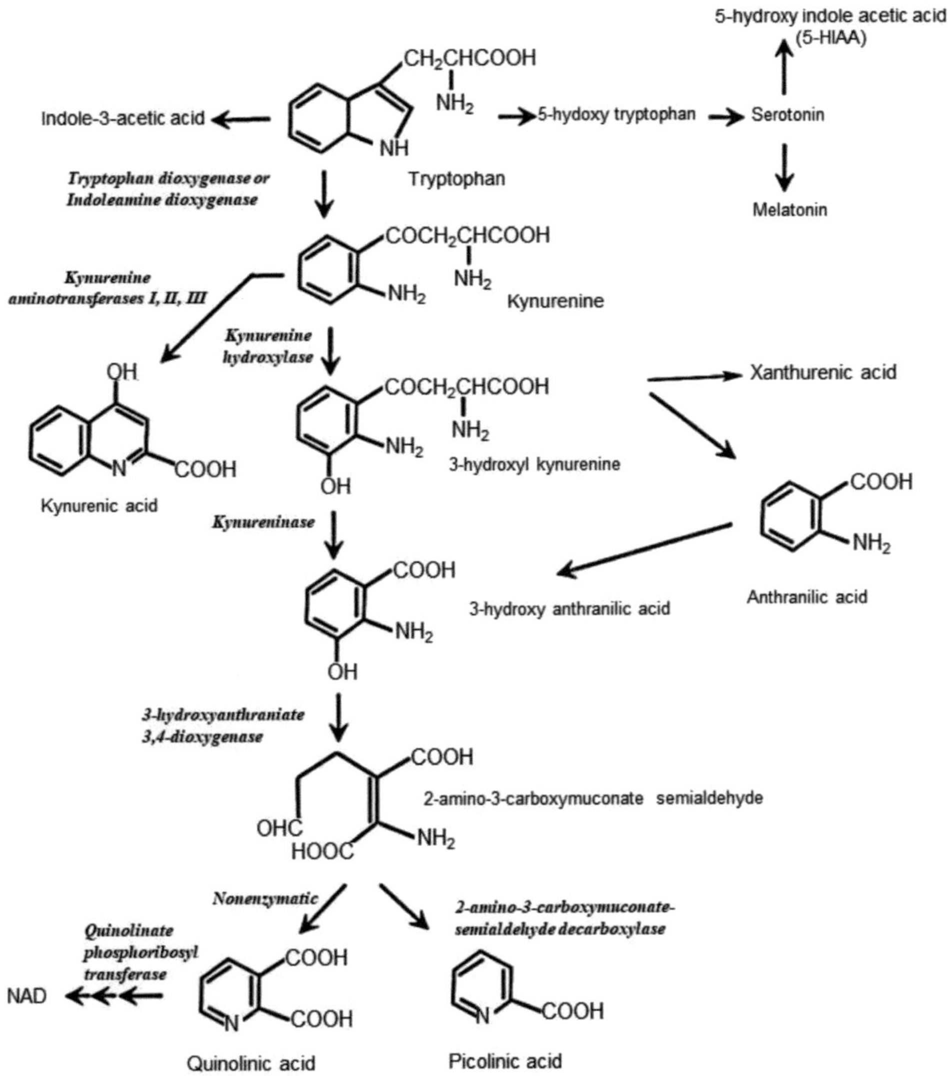

Figure 1.
Metabolic pathways of tryptophan.

Generally speaking, plasma levels of 5-hydroxyindoleacetic acid (5-HIAA), indole-3-acetic acid (IAA), KYN, and AA are higher in young women and old men than in young men and old women. Plasma levels of XA and 3HK are higher in young men and old women than in young women and old men.

One-way ANOVA was used for evaluating statistical significance. A, b, c, and d indicate values of young and old men and women. Tukey's test was used for post hoc test.

Table 3 shows that in contrast to cases in healthy people, age and gender differences disappeared in MMD except for KYN.

Sample name	young men n=25 a	young women n=10 b	old men n=20 c	old women n=20 d	significant differences
Tryptophan (μM)	63.92±10.83	67.79±9.30	68.89±13.32	47.68±7.19	a vs. d : p<0.01 b vs. d : p<0.01 c vs. d : p<0.01
L-5-Hydroxytryptophan (nM)	1.32±1.88	0.00±0.00	0.63±1.32	0.56±0.95	-
Serotonin (nM)	1.556±1.79	42.94±35.42	30.95±68.23	4.70±5.86	-
Kynurenine (nM)	693.9±177.8	1588±313	1696±1074	719.9±173.9	a vs. b : p<0.01 a vs. c : p<0.01 b vs. d : p<0.01 c vs. d : p<0.01
5-hydroxy-tryptophol (nM)	1.56±5.93	0.00±0.00	0.92±4.07	0.13±0.526	-
Tryptophol (nM)	0.09±0.38	4.96±6.59	1.19±2.95	0.10±0.42	-
5-Hydroxyindoleacetic acid (nM)	9.00±1.92	30.87±9.55	33.15±12.57	13.17±3.30	a vs. b : p<0.01 a vs. c : p<0.01 c vs. d : p<0.01
Indole-3-acetic acid (nM)	280.8±161.1	2446±1686	2433±854	262.8±142.4	a vs. b : p<0.01 a vs. c : p<0.01 b vs. d : p<0.01 c vs. d : p<0.01
Anthranilic acid (nM)	3.57±2.93	7.72±8.592	16.04±12.05	6.02±6.38	a vs. c : p<0.01 c vs. d : p<0.01
Kynurenic acid (nM)	73.64±23.74	53.20±16.66	64.03±21.65	65.40±26.12	-
Quinaldic acid (nM)	5.31±4.25	8.87±5.12	8.82±7.76	4.07±2.67	-
3-Indolebutyric acid (nM)	0.97±1.19	15.35±8.38	5.94±4.78	1.24±2.26	a vs. b : p<0.01 a vs. c : p<0.01 b vs. d : p<0.01 b vs. c : p<0.01 c vs. d : p<0.01
3-Hydroxykynurenine (nM)	10.31±2.24	4.42±1.27	2.60±2.24	12.80±3.80	a vs. b : p<0.01 a vs. c : p<0.01 a vs. d : p<0.05 b vs. d : p<0.01 c vs. d : p<0.01
3-hydroxyanthranilic acid (nM)	4.81±3.12	15.26±6.08	10.34±8.23	4.17±4.39	a vs. b : p<0.01 b vs. d : p<0.01
Xanthurenic acid (nM)	53.05±21.73	12.46±7.26	12.37±5.29	43.86±24.64	a vs. b : p<0.01 a vs. c : p<0.01 b vs. d : p<0.01 c vs. d : p<0.01

One-way ANOVA was used for evaluating statistical significance. A, b, c, d indicate values of young and old men and women. Tukey's test was used for post hoc test.

Table 2.
Measurements of plasma levels of TRP metabolites in healthy volunteers.

Diagnosis	MDD n=38				
Sample name	young men n=7 a	young women n=18 b	old men n=3 c	old women n=10 d	significant differences
Age (years)	36.00±9.256	40.00±7.004	60.00±14.73	59.40±7.306	a vs c: p<0.01 a vs d: p<0.01 b vs c: p<0.01 b vs d: p<0.01 b vs d: p<0.01
Height (m)	1.72±0.07	1.59±0.06	1.64±0.10	1.59±0.05	a vs b: p<0.01 a vs d: p<0.01
Weight (kg)	74.41±12.86	55.92±7.670	58.95±16.05	54.52±8.389	a vs b: p<0.01 a vs d: p<0.01
BMI	24.97±3.998	22.31±4.051	22.51±2.772	21.58±2.982	-
CGI-S	4.429±0.535	3.667±0.907	4.667±0.577	3.700±0.483	-
SDS	62.43±6.754	45.00±13.31	58.00±1.414	52.22±6.399	a vs b: p<0.01
HDR-S	21.29±3.817	16.94±5.150	29.67±5.033	18.30±4.398	b vs c: p<0.01 c vs d: p<0.01
Triptophan (μM)	64.09±13.77	58.48±10.40	59.11±2.026	51.24±11.01	-
L-5-Hydroxytryptophan (nM)	11.26±3.447	9.191±2.942	1.1800000	8.106±6.660	-
Serotonin (nM)	325.6±351.4	267.6±204.4	120.3±96.01	160.2±145.2	-
Kynurenine (nM)	2506.5±1042.9	1653.0±510.16	1014.2±418.25	1666.4±534.60	a vs b: p<0.05 a vs c: p<0.05
5-Hydroxytryptophol (nM)	-	10.50	25.11	2.20	-
Tryptophol (nM)	2.2950000	12.00±18.10	-	11.99±11.18	-
5-Hydroxyindoleacetic acid (nM)	28.04±22.07	35.00±28.99	22.82±28.11	48.68±58.22	-
Indole-3-acetic acid (nM)	2631.3±1666.0	2068.4±1525.8	583.0±266.4	2412.5±1026.1	-
Anthranilic acid (nM)	5.61	14.29±11.76	-	13.89±21.11	-
Kynurenic acid (nM)	67.07±39.75	32.09±15.37	26.47±12.76	36.57±14.38	-
Quinaldic acid (nM)	10.14±7.55	7.46±3.27	-	7.81±3.96	-
3-Indolebutyric acid (nM)	18.93±12.54	18.45±11.93	2.92±4.04	25.62±20.27	-
3-Hydroxykynurenine (nM)	19.03±14.17	20.55±11.06	29.96±8.09	13.98±5.98	-
3-hydroxyanthranilic acid (nM)	27.37±11.08	28.12±21.21	-	26.85±14.18	-
Xanthurenic acid (nM)	19.50±11.06	10.12±6.17	11.88±1.91	13.05±5.79	-
Melatonin (nM)	0.17	6.61±8.26	-	25.85±36.24	-
Quinolinic acid (positive) (nM)	776.1±1091.6	404.6±277.7	168.5±5.9	334.1±303.9	-

One-way ANOVA was used for evaluating statistical significance. A, b, c, d indicate values of young and old men and women. Tukey's test was used for post hoc test.

Table 3.
TRP metabolite levels of patients of MMD.

3. Discussion of part 1

The availability of endogenous 5-HT as a neurotransmitter is crucial in many physiological processes. Serotonergic neurons in the central nervous system are involved in regular behavioral states and physiological processes including arousal, sleep, appetite, pain, releases of hormone, and mood. Dysfunction of 5-HT neurons may lead to depression and other mental disorders.

Many scientific research has been done to know roles of 5-HT in pathophysiology of depression.

Since pathological changes are investigated in the brain and cerebrospinal fluid of suicides, it is claimed that 5-HT neurotransmission is implicated in the causes of suicide [13, 14]. Low levels of 5-HIAA in cerebrospinal fluid were shown in suicide attempters of depression [15]. Although the brainstem of suicide attempters had less 5-HT and 5-HIAA, most postmortem studies report no differences in cortical 5-HT or 5-HIAA of suicides [16].

Furthermore [17] patients with MDD have been reported to have higher 5-HIAA in jugular venous blood and have been argued to reflect higher brain 5-HT neurotransmission and turnover [18].

So the roles of 5-HT in depression is still confusing.

We simultaneously analyzed plasma levels of TRP metabolites in healthy people and patients of MMD. As shown in **Tables 2** and **3**, significant age and gender differences disappear in patients of MMD.

It is difficult to speculate reasons of such changes in MMD. Probably, hormonal changes may be implicated.

These results suggest that much attention has to be paid to age and gender if we want to analyze TRP metabolites, especially 5-HT and 5-HIAA.

Statistical differences of TRP metabolites between MMD or BD and healthy people will be reported elsewhere.

3.1 The diurnal variation of 5-HT in the blood of patients of depression

As stated above, serotonin (5-HT) plays roles in a state of depression since selective inhibitors of the uptake of 5-HT and the blockers of 5-HT 1A receptors are effective in its treatment [19, 20].

There is some evidence indicating that in patients of affective disorders, the regulation of circadian rhythms is disturbed [21].

We have shown that plasma levels of 5-HT were very low in patients of depression, but the levels of 5-HIAA or KYN were not different from the levels of control persons suggesting that 5-HT was immediately converted to 5-HIAA in patients of depression [9]. Due to the presence of 5-HT transporter in platelet membranes, most of 5-HT are believed to be stored in platelets in the blood [22].

We have shown that whole blood 5-HT concentration showed marked changes throughout daytime, with maximum values in the evening and lowest values in the morning, whereas its metabolite 5-HIAA followed contrary [23].

So we wanted to measure 5-HT levels in the blood of patients of depression and controls.

We examined at five timepoints whole blood 5-HT levels in depressive patients of Hamamatsu University Hospital and control volunteers. The number of depressive patients was 18 and 30 volunteers.

Patients were in depressive states as confirmed by a mean score of 18.7 (range 12–24) on the 24-item scale of Hamilton depression rating scale [24]. None of them were administered with any drug except for small doses of benzodiazepines, for at least 10 days before blood was taken.

Blood levels of 5-HT were measured using HPLC as described by Anderson et al [25]. Analytical recoveries were 85% (SD 4.5%, CV 5.6%). Amount and response were nearly related.

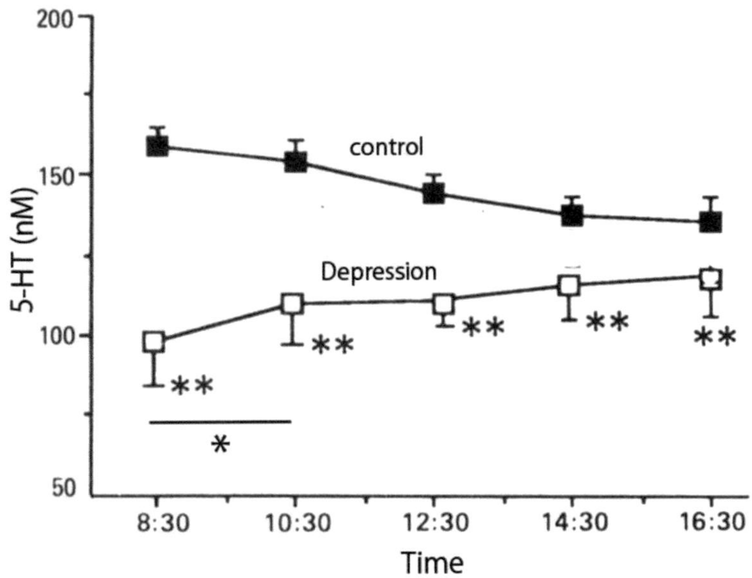

Figure 2.
Diurnal variation of blood serotonin levels of healthy men and patients of major depression. Ordinate, blood levels of 5-HT; abscissa, time when blood was taken. **$p < 0.01$ *control vs. depression,* *0.05, *levels at 8:30 vs. 10.30 of depressive patients.*

As shown in **Figure 2**, whole blood 5-HT levels were significantly lower in depressed patients at 8:30 10:30, 12:30, and 14:30. The blood concentration of 5-HT showed a circadian variation.

In the group of depression, the lowest value was shown at 8:30 and the level progressively increased to 14:30.

4. Discussion of part 2

Platelet 5-HT content is most likely regulated by the platelet transport activity.

Variations of 5-HT uptake in depressed patients have been reported by several groups [26–28].

Seasonal changes of serotonin (5-HT) uptake in blood platelets from depressed patients and normal controls were studied over a 2-year period to know if seasonal variations were present [26]. A measure of the number of 5-HT uptake sites in normal controls and depressed patients was significantly higher in fall and winter than in spring and summer. The number of 5-HT uptake in the depressed patients was lower than in normal controls throughout the year. Normal controls showed lower number in April and June. A similar trend was present in the depressed patients but the lowest values were found in the month of December.

Blood levels of melatonin, 5-HT, cortisol, prolactin, and serotonin uptake by platelets were measured at 08:00 to 08:00 hours of the following day in healthy men in age from 27 to 35 years [27]. The active transport of 5-HT by platelets was shown to be significantly correlated with melatonin blood levels. This finding suggests either a direct effect of melatonin on 5-HT active transport or the influence of the suprachiasmatic nucleus on serotonin uptake by platelets.

So far depressive disorders are considered to be associated with various neurobiological alterations like hyperactivity of the hypothalamic-pituitary-adrenal axis, altered neuroplasticity, and altered circadian rhythms. Unfortunately, the causal connections between depressive disorders and disturbed circadian rhythms have not been completely clarified. Chronobiological therapy is based on these disturbed processes. For the treatment of the circadian symptoms, various scientifically tested chronotherapeutics are available with different effectiveness and evidence like light therapy or sleep deprivation. The successful treatment of depression also frequently leads to an improvement in altered circadian rhythm.

Further studies of circadian variation of 5-HT system may help to understand the control of serotonergic nervous system and the treatment of depression.

Abbreviations

TRP	tryptophan
5-HT	serotonin
5-HIAA	5-hydroxyindoleacetic acid
IAA	indole-3-acetic acid
KYN	kynurenine
XA	xanthurenic acid
AA	anthranilic acid
KNA	kynurenic acid
3-HKN	3-hydroxykynurenine
IDO	indoleamine dioxygenase
TDO	tryptophan dioxygenase
SSRI	selective serotonin uptake inhibitor
SNRI	serotonin epinephrine reuptake inhibitor
CGI-S	clinical global impression—severity scale
SDS	self-rating depression scale
HDRS	Hamilton depression rating scale

Author details

Hiroi Tomioka[1,2], Junichi Masuda[3], Akikazu Takada[4*] and Akira Iwanami[1]

1 Department of Psychiatry, School of Medicine, Showa University, Japan

2 Department of Psychiatry, Northern Yokohama Hospital, Showa University, Japan

3 Global Application Development Center, Shimadzu Corporation, Kanagawa-ken, Japan

4 International Projects on Food and Health (NPO), Tokyo, Japan

*Address all correspondence to: takadaa@mwd.biglobe.ne.jp

References

[1] Walsh BT, Seidman SN, Sysko R, et al. Placebo response in studies of major depression: Variable, substantial, and growing. JAMA. 2002;**287**:1840-1847

[2] Kirsch I, Deacon BJ, Huedo-Medina TB, et al. Initial severity and antidepressant benefits: A meta-analysis of data submitted to the Food and Drug Administration. PLoS Medicine. 2008;**5**:e45

[3] Kirsch I. Clinical trial methodology and drug-placebo differences. World Psychiatry. 2015;**14**(3):301-302. DOI: 10.1002/wps.20242

[4] Cipriani A, Furukawa TA, Salanti G, Chaimani A, Atkinson LZ, Ogawa Y, et al. Comparative efficacy and acceptability of 21 antidepressant drugs for the acute treatment of adults with major depressive disorder: A systematic review and network meta-analysis. Lancet. 2018;**391**(10128):1357-1366

[5] Takada A, Curzon G. Serotonin in the central nervous system and periphery. In: International Congress Series 1088. Amsterdam, The Netherland: Excerpta Medica; 1995

[6] Chen Y, Guillemin GJ. Kynurenine pathway metabolites in humans: Disease and healthy states. International Journal of Tryptophan Research. 2009;**2**:1-19

[7] Mann JJ, Arango V, Marzuk PM, Theccanat S, Reis DJ. Evidence for the 5-HT hypothesis of suicide. A review of postmortem studies. The British Journal of Psychiatry. Supplement. 1989;**155**(Suppl 8):7-14

[8] Oquendo MA, Sullivan GM, Sudol K, Baca-Garcia E, Stanley BH, et al. Toward a biosignature for suicide. The American Journal of Psychiatry. 2014;**171**:1259-1277

[9] Takada A, Shimizu F, Masuda J. Measurement of plasma tryptophan metabolites: Clinical and experimental application for depression and stress states assessment. In: Melatonin—Molecular Biology, Clinical and Pharmaceutical Approaches. London, UK: Cristina Manuela Dragoi; 2018. DOI: 10.5772/intechopen.78560

[10] Shimizu F, Ishii Y, Ogawa M, Takao T, Matsuoka K, Kato K, et al. Plasma levels of tryptophan metabolites in healthy young and old men and women, and patients of type 2 diabetes mellitus (T2DM). Obesity: Open Access. 2018;**4**(2):1-4. DOI: 10.16966/2380-5528.138

[11] Takada A, Shimizu F, Masuda J. Plasma levels of tryptophan metabolites in patients of type 2 diabetes mellitus. In: Watson R, Preedy V, editors. Bioactive Food as Dietary Interventions for Diabetes. 2nd ed. MA, USA: Academic Press; 2019. pp. 265-276. DOI: 10.1016/B978-0-12-813822-9.00017-5

[12] Takada A, Shimizu F, Takao T, Masuda J. Measurement of tryptophan metabolites in healthy old men and patients of type 2 diabetes mellitus (T2DM). Food and Nutrition Sciences. 2018;**9**:1206-1220. DOI: 10.4236/fns.2018.910087

[13] Bach H, Arango V. In: Dwivedi Y, editor. Neuroanatomy of Serotonergic Abnormalities in Suicide. Boca Raton, FL: CRC Press/Taylor & Francis; 2012

[14] Mann JJ. Neurobiology of suicidal behaviour. Nature Reviews. Neuroscience. 2003;**4**:819-828

[15] Lidberg L, Belfrage H, Bertilsson L, Evenden MM, Åsberg M. Suicide attempts and impulse control disorder are related to low cerebrospinal

fluid 5-HIAA in mentally disordered violent offenders. Acta Psychiatrica Scandinavica. 2000;**101**:395-402

[16] Arango V, Mann JJ. Relevance of serotonergic postmortem studies to suicidal behavior. International Review of Psychiatry. 1992;**4**:131-140

[17] Andrews PW, Thomson JA Jr. The bright side of being blue: Depression as an adaptation for analyzing complex problems. Psychological Review. 2009;**116**:620-654

[18] Barton DA, Esler MD, Dawood T, Lambert EA, Haikerwal D, Brenchley C, et al. Elevated brain serotonin turnover in patients with depression: Effect of genotype and therapy. Archives of General Psychiatry. 2008;**65**:38-46. DOI: 10.1001/archgenpsychiatry.2007.11

[19] Jacobs BL, Fornal CA, Wilkinson LO. Neurophysiological and neurochemical studies of brain serotonergic neurons in behaving animals. Annals of the New York Academy of Sciences. 1990;**600**: 260-268. discussion 268-271

[20] Delgado PL, Charney DS, Price LH, Aghajanian GK, Landis H, Heninger GR. Serotonin function and the mechanism of antidepressant action. Reversal of antidepressant-induced remission by rapid depletion of plasma tryptophan. Archives of General Psychiatry. 1990;**47**:411-418

[21] Goodwin FK, Wirz-Justice A, Wehr TA. Evidence that the pathophysiology of depression and the mechanism of action of antidepressant drugs both involve alterations in circadian rhythms. Advances in Biochemical Psychopharmacology. 1982;**32**:1-11

[22] Pletscher A. The 5-hydroxytryptamine system of blood platelets: Physiology and pathophysiology. International Journal of Cardiology. 1987;**14**:177-188

[23] Pietraszek MH, Takahashi S, Takada Y, Ohara K, Inatomi H, Kondo N, et al. Diurnal patterns of serotonin, 5-hydroxyindoleacetic acid, tryptophan and fibrinolytic activity in blood of depressive patients and healthy volunteers. Thrombosis Research. 1991;**64**:243-252

[24] Hamilton M. Some aspects of the long-term treatment of severe hypertension with methyldopa. Postgraduate Medical Journal. 1968;**44**:66-69

[25] Anderson GM, Young JG, Cohen DJ, Schlicht KR, Patel N. Liquid-chromatographic determination of serotonin and tryptophan in whole blood and plasma. Clinical Chemistry. 1981;**27**:775-776

[26] Humphries LL, Shirley P, Allen M, Codd EE, Walker RF. Daily patterns of serotonin uptake in platelets from psychiatric patients and control volunteers. Biological Psychiatry. 1985;**20**:1073-1081

[27] Rausch JL, Shah NS, Burch EA, Donald AG. Platelet serotonin uptake in depressed patients: Circadian effect. Biological Psychiatry. 1982;**17**:121-123

[28] Modai I, Malmgren R, Wetterberg L, Enervoth P, Valevski A, Asberg M. Blood levels of melatonin, serotonin, cortisol, and prolactin in relation to the circadian rhythm of platelet serotonin uptake. Psychiatry Research. 1992;**43**:161-166